Tape Recorders—How They Work

by

Charles G. Westcott and Richard F. Dubbe

Revised by

Norman H. Crowhurst

HOWARD W. SAMS & CO., INC.
THE BOBBS-MERRILL CO., INC.
INDIANAPOLIS · KANSAS CITY · NEW YORK

THIRD EDITION

FIRST PRINTING—1974

International Standard Book Number: 0-672-20989-6
Library of Congress Catalog Card Number: 73-90290

Preface

The techniques and equipment used in tape recording have improved tremendously since the original edition of *Tape Recorders—How They Work* was published. This revised edition has been updated to include the latest in stereo and quadraphonic (four channel) recording equipment, as well as to exactly explain how tape recorders work.

Although it is possible to make good recordings without understanding the mechanics and electronics of the recorder, understanding the principles of your machine makes good recordings more certain and more frequent. Thus, the complete operation of the tape recorder is presented in a nontechnical language, starting with a short history of magnetic recording. Motorboards, record/playback heads, volume indicators, bias circuits and theory, tape-transport mechanisms, equalization circuits, recording and playback amplifiers, and much more are thoroughly explained. In addition, chapters on testing recorders and on special applications such as "sound-on-sound" recording, student, recorders, and disc recorders are included.

It is hoped that readers will find this new edition not only a valuable guide to tape recorder operation, but also a helpful means of increasing their enjoyment of the tape recorder.

Contents

CHAPTER 1

History of the Magnetic Recording Industry

The history of magnetic recording covers a period spanning more than half a century. Yet magnetic recording remained practically undeveloped until shortly before the beginning of World War II. From its inception in 1898 until its real commercial application in Germany starting in 1935, magnetic recording remained little more than a scientific curiosity. Historians of invention during that era regarded it as scarcely worth mentioning.

Valdemar Poulsen of Copenhagen, Denmark, pictured in Fig. 1-1, built the first magnetic recorder in 1898. His *Telegraphone* magnetically recorded crosswise on a steel piano wire, but the wire would twist and throw the crosswise recording out of alignment, ruining the quality. To overcome this difficulty, Poulsen replaced the wire with a more bulky and cumbersome steel tape that could not twist. A *Telegraphone* manufactured in the United States by the American Telegraphone Company around 1920 is shown in Fig. 1-2. The spools of wire are mounted horizontally and turned by the 100-volt motor pictured at the lower right. No provision for automatic rewind of the wire was made; the spools of wire had to be interchanged for rewind and playback.

In his United States patent granted on the *Telegraphone* (Fig. 1-3), Poulsen called it an "apparatus for electromagnetically receiving, recording, reproducing, and distributing articulate speech."

Poulsen's invention was unique and completely original, not the outgrowth or further refinement of a fundamental discovery made earlier. Recorded scientific history does not disclose or even give a possible clue as to where Poulsen got his idea. Although historians of invention passed Poulsen's discovery by with scarcely a nodding mention, the *Telegraphone* was nevertheless a sensation in its day. In 1900 at the Paris Exposition, the *Telegraphone* won the Grand Prix. Technical journals and newspapers were filled with references about it.

Fig. 1-1. Valdemar Poulsen, inventor of magnetic recording in 1898.

The *Telegraphone* particularly captured the imagination of investors in the United States. Poulsen was able to raise with little effort the princely sum of $5,000,000 in capital stock at a par value of $10 a share for the manufacture of the *Telegraphone.* The backers visualized its chief use as an office dictating device. The machine was capable of recording for 30 minutes on a wire traveling at an amazing velocity of 7 feet per second. The machine required earphones; it operated without an electronic amplifier—not destined to be discovered until years later.

A series of financial misfortunes beset Poulsen's United States firm, the American Telegraphone Company. Lacking present-day persuasive promotion and selling techniques, coupled with inherent defects in the recorder itself, the firm soon collapsed and the stock became worthless. A Danish firm was also organized in 1909 to manufacture the *Telegraphone.* The firm failed in 1916 without having commercially marketed a single recorder.

Electronics engineers who examine Poulsen's patent today are amazed at how closely his machine, crude as it was, resembles the

Courtesy Popular Electronics

Fig. 1-2. The *Telegraphone* designed by Poulsen and manufactured by the American Telegraphone Company around 1920.

modern magnetic recorder. Oscar Dupue, in an account of his experiences with Burton Holmes, then the world's leading travelogue exponent, recalled having used the *Telegraphone*:*

> "I have mentioned previously the second trip to Norway in 1907 to make another film of the fjord trips. It was on this trip that I purchased a Poulsen wire recorder in Copenhagen . . . I was able to operate it in my steamer cabin while en route home. I had a lot of fun talking into it and playing back, and soon I had a procession of passengers eager to record and hear their own voices. Several theatrical notables were present, including the famous Jimmie Powers . . . he was full of hit songs and stories, so we recorded a few. When he finished, I spoke

* Oscar Dupue, "My First Fifty Years in Motion Pictures," *Journal of the Society of Motion Picture Engineers,* December, 1947.

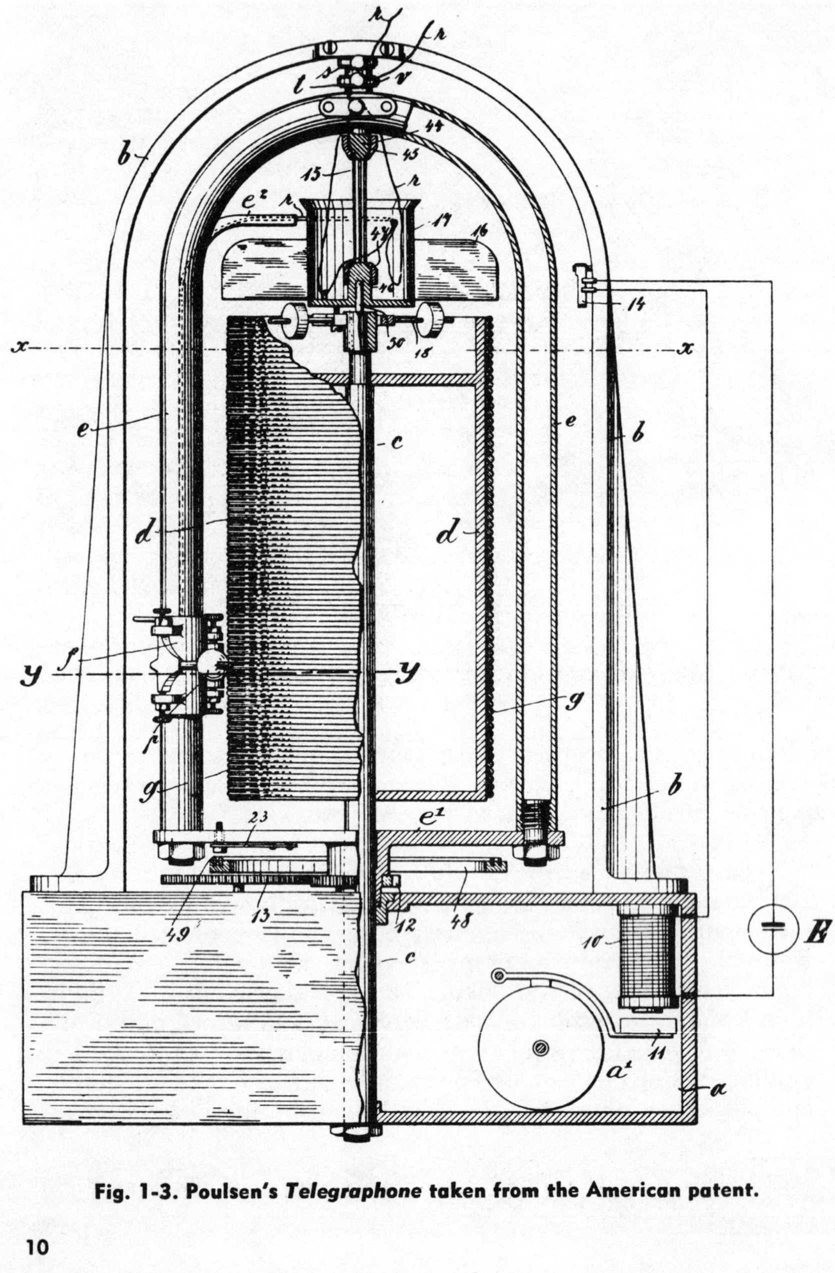

Fig. 1-3. Poulsen's *Telegraphone* taken from the American patent.

> into the recorder saying that Powers' record was made on the twenty-eighth day of August, 1907, in midocean aboard the S.S. *Augusta Victoria*. Thirty years later, I rerecorded Powers' voice on film. The wire had retained the record as clearly as when it was first made."

Since Poulsen had a working magnetic recorder fifty years ago, why did magnetic recording have to wait until World War II to be fully developed? One obvious answer to this question lies in Poulsen's recorder. Since the recorder originally employed wire and later a steel tape, it was undoubtedly very awkward and difficult to operate. Because the speed at which the tape traveled in recording was as fast as was practical for the mechanism to handle, it was impossible to speed up the rewind time. Also, since it recorded crosswise or in a perpendicular direction, the quality could not have been good, although undoubtedly it represented an achievement when contrasted to the sound standard prevailing at the time. Frequency response was limited. The dynamic recording range did not exceed approximately 20 decibels. The noise level was excessively high. However, the main reason for the lack of development can be attributed to the low acoustical output of magnetic recording in comparison with the then competitive mechanical systems such as disc recordings.

Not until the late Dr. Lee de Forest, pictured in Fig. 1-4, invented the triode vacuum tube was it possible to obtain sufficient amplification to satisfactorily reproduce the weak signal on magnetic tape. In fact, the relative positions of the now fast rising magnetic recording medium and phonograph industry might today have been reversed through a little known experiment conducted in 1912. But as is so frequent in the history of invention, the full significance of the experiment was not realized at the time.

Dr. Lee de Forest wrote:

> "Speaking of the necessity for the three-electrode tube amplifier in connection with magnetic-tape recording and reproducing, you may be interested to know that in the spring of 1912 I used the tube as an amplifier in connection with the old steel-wire Poulsen *Telegraphone*. I am sure this is the first combination of those two great inventions. No one could foresee at that early date the immense development and priceless applications of the magnetic-tape *Telegraphone*. In fact, one of the very first applications of the three-electrode tube as an amplifier was in connection with the *Telegraphone*."

Other roadblocks besides amplification existed. The two principal ones were removed with the abandonment of perpendicular magnetization in favor of longitudinal magnetization and the discovery of the ac or high-frequency bias technique. Originally the magnetic field was recorded perpendicular to the recording medium (wire, steel band, or tape). While this system worked, it required very high tape speeds since the magnetic field in the region of the tape could not be concentrated but covered a wide area (Fig. 1-5).

However, with the invention of the ring-type head, the magnetic field could be confined, as far as the tape is concerned, to the area between the pole pieces or to the magnetic gap. This permitted recording wavelengths as short as 0.00025 inch (video-recording wavelengths can now be recorded to a fraction of this amount).

Thus, with the ring-type head and longitudinal recording (recording in the direction of tape travel rather than perpendicular to it as illustrated in Fig. 1-6), engineers were able to make low tape speeds and the storing of a large amount of information on a small reel economically practical.

The ac or high-frequency bias technique was discovered by W. L. Carlson and G. W. Carpenter of the U.S. Navy. The ac bias technique eliminated the high background noise generally associated with recordings made using the older dc bias method. However, even with the removal of these obstacles, the progress of magnetic recordings was slowed for many years through lack of imagination and any serious desire to perfect existing techniques.

Fig. 1-4. Dr. Lee de Forest, inventor of the triode vacuum tube.

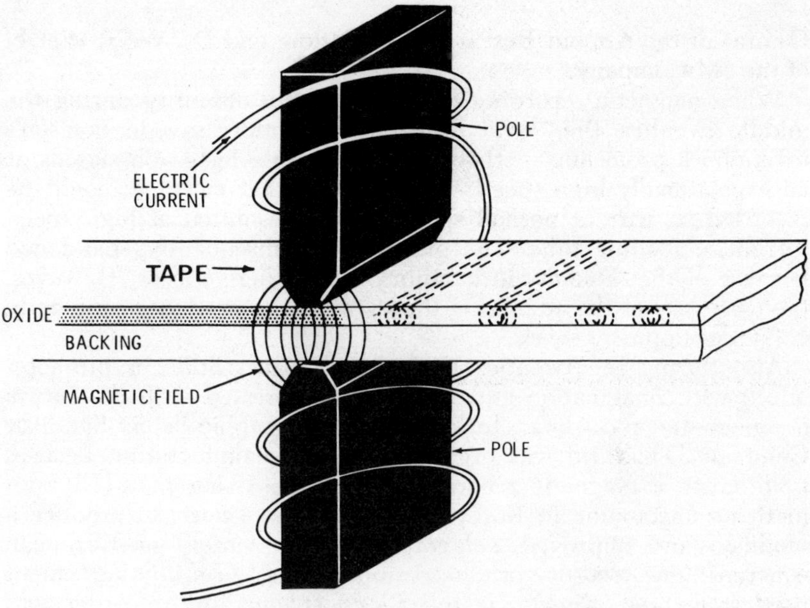

Fig. 1-5. Perpendicular magnetic recording.

The present state of development is largely due to the early work of such leaders in the field as Allgemeine Electrizitats Gesellschaft (AEG) in Germany, manufacturers of the *Magnetophone;* the Bell Telephone Laboratories; Lynn Holmes of the Stromberg-Carlson Company; S. J. Begun of the Brush Development Company; Marvin

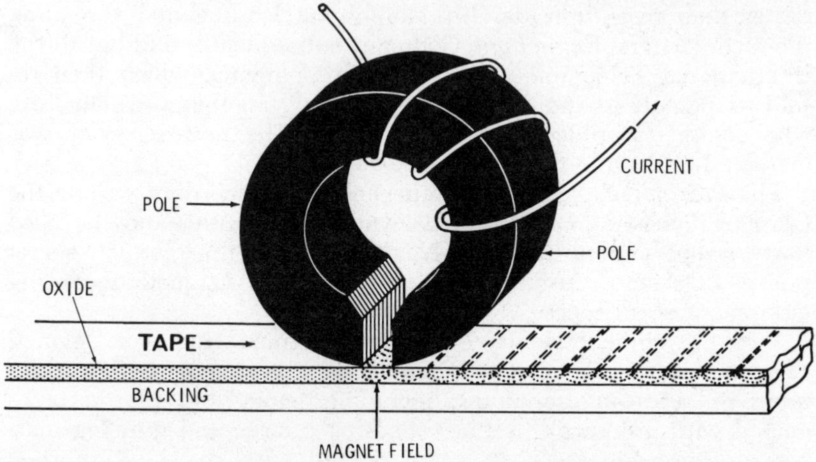

Fig. 1-6. Longitudinal magnetic recording.

Camras of the Armour Research Foundation; and Dr. W.W. Wetzel of the 3M Company.

When magnetic recording bobbed up from obscurity during the middle twenties, the Naval Research Laboratory saw in Poulsen's invention a promising method for transmitting telegraph signals at an exceptionally high speed. It was hoped that messages could be recorded on wire at normal speed, then transmitted at high speed to reduce sending time. The project was subsequently abandoned because of the difficulty in handling wire at high speeds. However, the discovery of the ac biasing theory was a by-product of the early Navy development work.

Also during the twenties in Germany, Kurt Stille, a promoter gifted with imagination and vision and impressed with the future of magnetic recording, formed the Telegraphic-Patent-Syndikat Company. This firm was organized to sell manufacturing licenses to produce a magnetic recorder/reproducer. Using hard-sell sales methods uncommon in Europe, Stille sold his license to produce a modified and improved *Telegraphone* to a variety of European concerns. One recorder produced under Stille's licensing agreement used steel tape. Known as the *Blattnerphone,* this recorder was seriously considered for early talking motion pictures. In England, several movies were completed that actually used a sound track recorded on synchronized steel tape.

In Germany particularly, a variety of magnetic recorders were marketed during the early thirties, all operating with either wire or steel bands. The *Echophone* Company, to whom Kurt Stille sold a license agreement, was one of the more prominent German concerns. Their recorder was the first to use a magazine-load device rather than reels, considerably simplifying loading and threading the recorder. The *Echophone* Company subsequently sold out to the International Telephone and Telegraph Company, which then resold its patents to the C. Lorenz Company, another German firm, who completely redesigned the machine. The new recorder was marketed under the trade name *Textophone.*

The *Textophone,* along with other magnetic recorders, was on the German domestic market in 1933 when Adolf Hitler and the Nazi Party swept into power. The Nazi Party and the German secret police (Gestapo) used large quantities of magnetic-recording equipment of all types.

Working independently, another German inventor named Pfleumer was busily engaged conducting experiments with various types of recording mediums, including paper and plastic tapes coated with iron-oxide particles. Sensing a large and growing market for magnetic recorders, AEG joined with the I. G. Farben Company and took over Pfleumer's early pioneering work in coating paper and plastic materials. The grain size of the magnetic materials was relatively large. The early paper tape, coated with the powdery magnetic substance, closely resembled sandpaper. When the tape was run through a recorder, sprays of powder clouded the air.

The first *Magnetophone,* produced by AEG, was exhibited in 1935 at the German Annual Radio Exposition in Berlin. Although the *Magnetophone* was inferior to many earlier German magnetic recorders, it was an instantaneous success and the hit of the exhibition. Employing coated paper tape as opposed to wire or steel bands, the magnetic-recording tape cost only fifteen cents per minute of recording time; the price of steel bands was a dollar or more per minute. Naturally, paper tape was less unwieldy, threaded easily, and was infinitely more convenient to store.

Meanwhile, in the United States, Bell Telephone Laboratories was the only large corporate name in the electronics field to be associated with magnetic recording. Bell Telephone Laboratories designed a steel-tape recorder, the *Mirrophone,* that was used to report weather and time information over the telephone. At the New York World's Fair in 1939, the *Mirrophone* amazed thousands of visitors who gasped in surprise to hear a chance remark of a moment instantly played back with full fidelity. the *Mirrophone* was also featured in the Bell Telephone exhibits which toured the schools of the nation as part of a continuing public relations program.

Magnetic-recording equipment had not been manufactured commercially in the United States until 1937 when the *Soundmirror,* a steel-tape recorder, was placed on the market. Brush Development Company purchased rights to the *Soundmirror,* which was originally built by Acoustic Consultants. Fig. 1-7 shows one of the early Brush *Soundmirror* recorders. The cover is removed so that the mechanical layout may be seen. Although the machine could record only one minute on a steel tape, countless applications were immediately found for it. Brush inaugurated an intensive research program which made many fundamental contributions to the body of knowledge comprising magnetic-recording techniques. During World War II, Brush built large quantities of magnetic-recording equipment for the Armed Services. In cooperation with the Office of Scientific Research and Development, Brush helped develop coated paper-tape and plated-wire recording mediums.

Armour Research Foundation was also prominent in the early development of magnetic recording. The chief contribution of the Armour Research Foundation lies in its intensive campaign launched to acquaint the public with the full potentialities of magnetic recording. It also supplied magnetic-recording equipment to the Armed Services during World War II. However, because of limited shop facilities at Armour, General Electric and several other companies took over production of the Armour Research Foundation's recorder.

Immediately following the end of World War II, Webster-Chicago and Sears, Roebuck and Company began large-scale wire-recorder production. War workers with pockets still filled with money provided a wide home consumer market base, and businesses who bought recorders for specialized office dictation uses and other applications topped off the market. Encouragingly large quantities of wire recorders were sold during the first few months of large-scale production. However, commercial television began to show promise of sweeping the country, and with automatic home appliances, automobiles, and other types of consumer products increasingly available, sales of wire recorders began to slow. The anticipated wire-recording boom failed to materialize.

During World War II, all avenues of exchanging technical information between Germany and the outside world were closed.

Fig. 1-7. An early *Soundmirror* recorder with cover removed.

But work on the magnetic-recording process progressed, despite shortages and hardships imposed by the war. Substantial improvements were being made on the AEG *Magnetophone.*

Near the close of World War II, a corps of Allied scientists and technicians was sent to Germany to investigate reports of a highly developed magnetic-tape and tape-recorder system for general recording purposes.

American technicians were amazed to find that the *Magnetophone,* employing ac bias and using a coated paper tape, was able to record a full 10,000 cycles with a low signal-to-noise ratio and negligible wow and flutter. Although experimental American wire recorders under carefully controlled conditions could now attain 10,000 cycles, wire had many defects when compared to tape. Quick to see the advantages of the tape system over other recording methods, the investigators returned prototypes of the recorder and samples of the tape to the United States. Immediately an intensive developmental program was begun, aimed at opening up the field of magnetic recording in the United States.

The United States Alien Property Custodian held all patents on the *Magnetophone* and would license any American firm to produce the equipment on application. Although any company that wished to duplicate the *Magnetophone* could do so, large corporations still showed little interest. For the first few years following the war, practically all manufacturing of professional recorders was done by three small and then little known organizations: Magnecord, Inc., Rangertone, Inc., and Ampex Electric Company. Samples of early units manufactured by these companies are shown in Fig. 1-8.

While the early American recorders were directly copied from the German AEG *Magnetophone,* the German magnetic tape was found to be far from satisfactory.

The early German tapes used a type of oxide known as *raven red,* a term borrowed from the barn paint of the same name. Literally a high grade of barn paint, the oxide was painted on a paper backing. The major deficiencies of the German tape were three in number. First, the available output from German tape was very low and required large amplifiers in the recorders for satisfactory performance. Second, the tape had poor response at short wavelengths, and this necessitated driving the tape at a high velocity to maintain adequate frequency response. Finally, the uniformity of the German tape, lot to lot and even within a single reel, varied widely. This nonuniformity made standardization of recording and playback equipment and the recording procedure itself difficult.

Because of the limitations imposed by their tape, the Germans had arbitrarily selected a high speed that would give good frequency response within practical limits. The first American ma-

(*A*) *Magnecord PT6-A.*

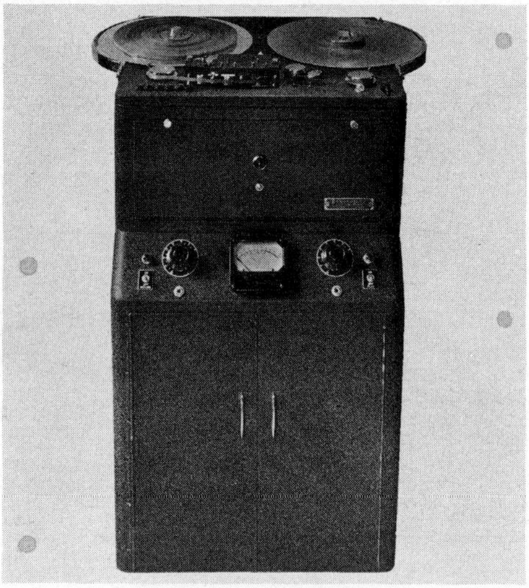

(*B*) *Rangertone.*

(*C*) *Ampex.*

Fig. 1-8. Early professional-type tape recorders.

chines were built with a tape speed as close as possible to the German speed of 76 centimeters per second. Thus, the speed of the first American recorder was 30 inches per second.

Early attempts at manufacturing magnetic tape were made by the Brush Development Company and Indiana Steel, who used a unique magnetic-iron material with high coercive force but which was difficult to record and erase. Minnesota Mining and Manufacturing Company, in mastering the production of high quality magnetic tape, made substantial improvements over the early German attempts. The task was not an easy one. The technical specifications for the production of quality magnetic tape were far more exacting than those for any other type of coating process previously known.

The problem of high tape speed was solved by the introduction of a new type of magnetic oxide. The new oxide gave the American tape about 12-dB more output than was available with the German tape. The higher coercivity of the oxide gave the tape excellent high-frequency response characteristics, making possible the adoption of slower tape speeds.

With the improvements made in magnetic tape, recorder speeds were progressively halved, first from 30 to 15 inches per second (ips), then from 15 ips to 7.5, 3.75, 1⅞, and 15/16 ips. Table 1-1 gives

Table 1-1. Recording Time for Various Magnetic-Tape Speeds and Tape Lengths

Reel Size	Tape Length	Speed				
		15/16 ips	1 7/8 ips	3 3/4 ips	7 1/2 ips	15 ips
3"	150'	30 min	15 min	7 1/2 min	3 3/4 min	1 7/8 min
3"	*225'	45 min	22 1/2 min	11 1/4 min	5 5/8 min	2 13/16 min
4"	300'	1 hour	30 min	15 min	7 1/2 min	3 3/4 min
5"	600'	2 hours	1 hour	30 min	15 min	7 1/2 min
5"	*900'	3 hours	90 min	45 min	22 1/2 min	11 1/4 min
5"	†1200'	4 hours	20 hours	1 hour	30 min	15 min
5"	‡1800'	6 hours	3 hours	90 min	45 min	22 1/2 min
7"	1200'	4 hours	2 hours	1 hour	30 min	15 min
7"	1500'	5 1/3 hours	2 2/3 hours	80 min	40 min	20 min
7"	*1800'	6 hours	3 hours	90 min	45 min	22 1/2 min
7"	2000'	7 hours	3 1/2 hours	106 min	53 min	27 min
7"	+2400'	8 hours	4 hours	2 hours	60 min	30 min
7"	3000'	10 2/3 hours	5 2/3 hours	2 2/3 hours	80 min	40 min
7"	‡3600'	12 hours	6 hours	3 hours	90 min	45 min
10 1/2"	2400'	8 hours	4 hours	2 hours	1 hour	30 min
10 1/2"	*3600'	12 hours	6 hours	3 hours	1 1/2 hours	45 min
10 1/2"	†4800'	16 hours	8 hours	4 hours	2 hours	60 min
14"	5000'	16 1/2 hours	8 1/3 hours	4 1/6 hours	125 min	62 1/2 min
14"	*7200'	24 hours	12 hours	6 hours	3 hours	90 min

* 1-mil extra-recording tape.
† 1/2-mil double-recording tape.
‡ Tempered-mylar triple-recording tape.
All times are for 1-direction 1-track only.
For total mono time on multiple-track recorders, multiply by number of tracks.
For total stereo time on multiple-track recorders, multiply by one-half the number of tracks.

the recording time obtainable at each speed for various size reels. Since it is easier to build machines that work in multiple speeds of each other, each progressive speed change was obtained by dividing by two.

Tape speed is one important factor in determining the upper frequency limit of a recorder. Until recently it was possible to obtain one thousand cycles of response for every inch of tape speed using an arbitrary rule-of-thumb measurement. For example, if a recorder operated at a tape speed of 15 ips, it was possible to get a 15,000-hertz response. However, recent improvements in magnetic heads, recorder design, and still further advances in magnetic-tape construction have permitted the successive extension of the original capabilities many times over. With the addition of noise-reduction electronics, performance at 1⅞ ips easily equals former response at 15 ips.

The earliest heads used high-permeability, low-loss laminated materials similar to those used for high-quality input transformers used with microphones. Production techniques and the introduction of better magnetic materials for this purpose made possible the manufacture of heads capable of handling finer tape materials.

These improvements progressed concurrently. While one company was busy developing new oxide materials with higher coercivity and capable of accepting a higher level of magnetization, other companies were perfecting head designs and methods of production. Improvement in uniformity allowed the compression of more tracks into the same tape width, which for many years had remained at the ¼-inch standard. A variety of superhigh-quality systems were developed for professional recordings, particularly those used as masters for preserving tape or disc originals from which production records are made.

Typical tape widths of ½ inch, 1 inch, and 35 millimeters with various numbers of tracks were used. The extra tracks did not provide extra playing time as they do on the marketed version multiple tracks. Instead, they enable the original master to be recorded in many more parts—a sort of superstereo system.

This gave much greater control in assembling the final release version. The original performance could be reproduced; then the reproduction could be altered for the final print, either mono or stereo. Endless variations were possible until both the artist and the studio personnel were satisfied that they had a good "commercial" product.

Another method for compressing more playing time into the same space was to reduce the thickness of the magnetic tape, thereby permitting the use of a thinner magnetic tape than previously possible. At first, the elastic quality of the thin plastic tape caused it to stretch during handling, resulting in mechanical distortions that ruined the recording. Later on, the introduction of more stable plastics, particularly Mylar, remedied that problem.

However, the thinness of the tape caused another problem; each layer of magnetic coating was so close to the next one that the two were superimposed on each other. This exaggerated imperfections in the tape and caused print-through problems. Print-through occurs when the magnetic field from one layer of tape somewhat magnetizes the adjacent layers, causing a signal transfer or a faint reproduction of sound from one layer of tape to appear as background echo on the next layer. The development of tape with a greater coercive force reduced the print-through, but the real breakthrough came with the introduction of chromium oxide, a considerable improvement over the iron oxides of the early days. The use of chromium oxide allowed the taping of higher intensity recordings, thus almost eliminating print-through.

Credit for refinements of the tape recorder must also be shared with electronic advancements that decreased the space necessary to handle the tape and circuitry systems. Fig. 1-9 shows an early professional installation; a lot of space was needed to process the tape as well as house the electronic system. In contrast, Fig. 1-10 shows an approximately equivalent modern system. Multitrack tapes plus solid-state circuitry have increased the capabilities of the system and minimized space requirements.

Fig. 1-9. A typical professional recording installation of a few years ago.

Other changes overcame the advantages of the disc over the tape as recording media. The disc was simple to play—the stylus could be inserted in the record groove manually by hand or automatically by the record changer mechanism. Multiple copies were rapidly produced from the disc, and its form was easy to handle. However, the tape was more versatile—it could record, erase, and rerecord on multiple tracks to obtain a quality recording. But the tape was hard to handle and reproduction was a laborious process.

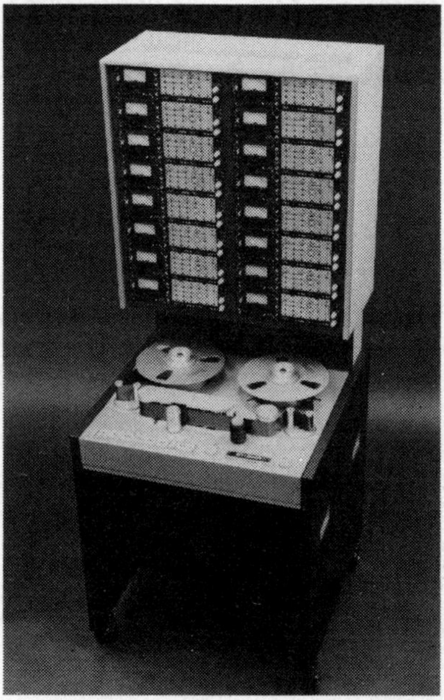

Courtesy Gotham Audio

Fig. 1-10. An imported portable system with greater capability than the much larger one of Fig. 1-9.

The advent of the cassette clearly established the supremacy of the tape over the disc. The cassette made it easy to load and play tapes, nor was there a turntable that had to be kept level and stationary. Copying could be performed at many times the recording speed, and the copies turned out at phenomenal rates.

The cassette is one of three types of prepackaged tapes on the market today; reel-to-reel and cartridge are the other two. Each must be used with a tape recorder designed for it.

The original reel-to-reel tape is usually hand-threaded onto the take-up reel although some sophisticated self-threading models are also available.

The cartridge tape is convenient for continuous playing, and it is easily inserted into the tape machine. Playing is accomplished by

Courtesy JVC

(A) *Original reel-to-reel used mostly for decks.*

Courtesy Hitachi

(B) *Cassette portable with built-in microphone.*

Courtesy Magnavox

(C) *Cartridge-type recorder deck.*

Fig. 1-11. Samples of three home tape recorders.

withdrawing the tape from the center of its spool, passing it through the playing heads and transport mechanism, and feeding it back onto the outside of the spool. When a track is finished, the tape comes back to its starting point, ready to play again. The disadvantage of the cartridge is that it is not reversible. Once a given spot on the track has been passed, it cannot be repeated until that track is replayed from the beginning. However, the cartridge can have multiple tracks and can be switched from one track to another at any point.

The cassette tape has two spools of tape which play two ways—half the tracks in one direction and half in the other direction. It can be inserted or removed at any spot. Like the cartridge, the cassette is simple to load and handle.

At the present time, numerous brands of magnetic tape and tape machines are marketed. This competitiveness accounts for the rapid technological advancements in the recording field, which have made the tape recorder a fundamental recording tool. We will discuss the application of some of these refinements in later chapters.

CHAPTER 2

Theory of Magnetic Recording

The first time Dad brings home a tape recorder to record the children's voices or practice his speech for the Company's annual sales meeting, the entire family gathers around. Following the initial flush of excitement and after everyone has recorded his voice, Dad is asked the inevitable question, "How does it work?"

Typically, Dad might reply, "Why, with push buttons, of course."

Yet an increasing number of tape recorder owners have not been content to be mere "button pushers" and as a result have started the still growing cult of amateur sound hunters and golden-ear addicts that has turned high fidelity into a challenging but rewarding hobby.

As the tape-recorder owner's enthusiasm increases, his ear slowly changes from tin to the golden variety. He will drive miles to compare his tapes with those of fellow amateurs or with the results of local professionals. As a result, he will invariably seek to improve his technique.

Just as the camera fan soon realizes that he can improve his pictures by learning what happens when he clicks the shutter, the tape enthusiast learns how to get the most out of his recorder when he understands how it works.

An abundance of papers has been presented before technical society symposiums on the theory and principles of magnetic recording. Technical journals repeatedly carry detailed scientific material, and books have been written on the subject of magnetic recording. Nevertheless, physicists and engineers are not in complete agreement on some of the finer points. Symposiums have broken up in disagreement. Controversy is still prevalent among magnetic-recording theoreticians, a healthy sign of industry growth and progress.

A thorough understanding of the technical phase of magnetic recording requires background knowledge and experience in electricity and magnetism. The basic principles are not difficult to understand; a grasp of the fundamentals of magnetic recording is basic to getting the best performance out of any tape recorder—from the simplest home machine to the finest studio recorder.

SOUND WAVES—ELECTRICAL WAVES

Sound, by its very nature, is an auditory sensation experienced through the ear. Sound is caused by an alteration in pressure, particle displacement or particle velocity produced when the air is set into motion (Fig. 2-1). The action of sound waves radiating from a single source can be compared to dropping a stone into a still pond. The waves of water move outward in all directions, continuously expanding as shown in Fig. 2-2. Sound waves, likewise, continuously expand outward from the sound source.

Sound may be produced by a vibrating body such as the human diaphragm, a violin string, or an intermittent air stream such as a human voice or a lip-reed instrument. Sound travels through the air in surges of energy, consisting of a compression or a high-pressure pulse of air followed by a rarefaction or a low-pressure pulse. Anyone who has stood at the seashore, fascinated with the ocean waves crashing against the beach, will note its similarity to sound waves. The ocean wave surges forward, draws back, and plunges forward once again.

The human ear receives the sound waves and converts them into electrical impulses which are "translated" by the nervous system of our body.

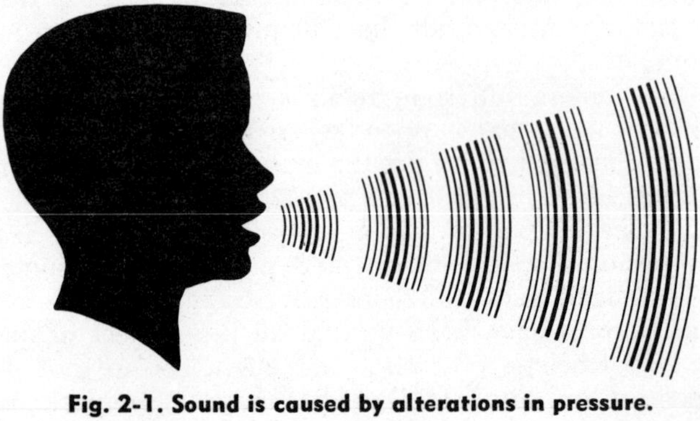

Fig. 2-1. Sound is caused by alterations in pressure.

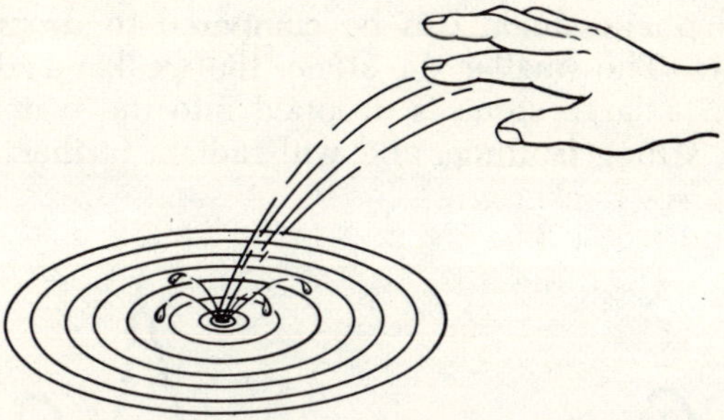

Fig. 2-2. Sound waves radiating from a source compared to dropping a stone into a still pond.

Sound waves, like electrical waves or impulses, can be graphed. Variations of pressure—concentration and rarefaction of the air particles—when plotted, form a wave. If the sound consists of only one frequency and is of constant strength, the wave will be sinusoidal in shape. Sound waves of a human voice or a musical instrument are composed of many frequencies, each with its own particular strength, all existing at one time.

Fig. 2-3 shows the amplitude of a sinusoidal wave and the number of times it repeats itself in a second. The height of the diagramed sound wave indicates the intensity or loudness of a sound, while the number of complete wave cycles per second (hertz) gives the pitch of the sound. The more wave cycles per second, the higher the pitch. Fig. 2-4 shows a complex waveform indicating a number of frequencies of different amplitudes occurring simultaneously.

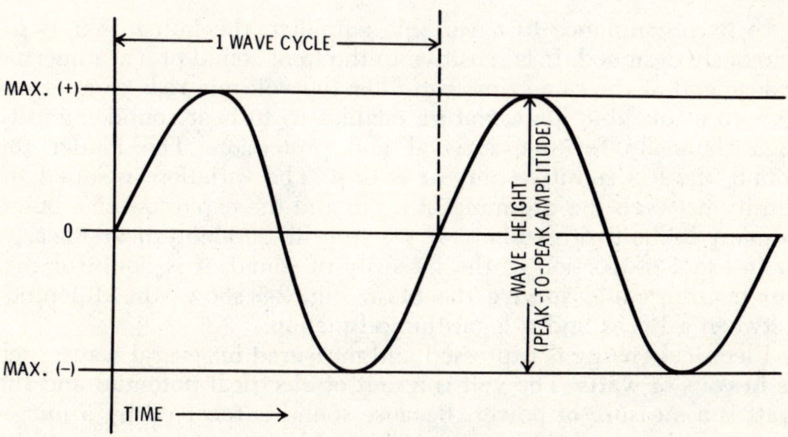

Fig. 2-3. A sinusoidal wave.

Again, this phenomenon can be compared to dropping a stone into still water. The smaller the stone, the weaker and smaller the ripples. When a large stone is dropped into the water, the waves will be large, strong in force, and will radiate farther.

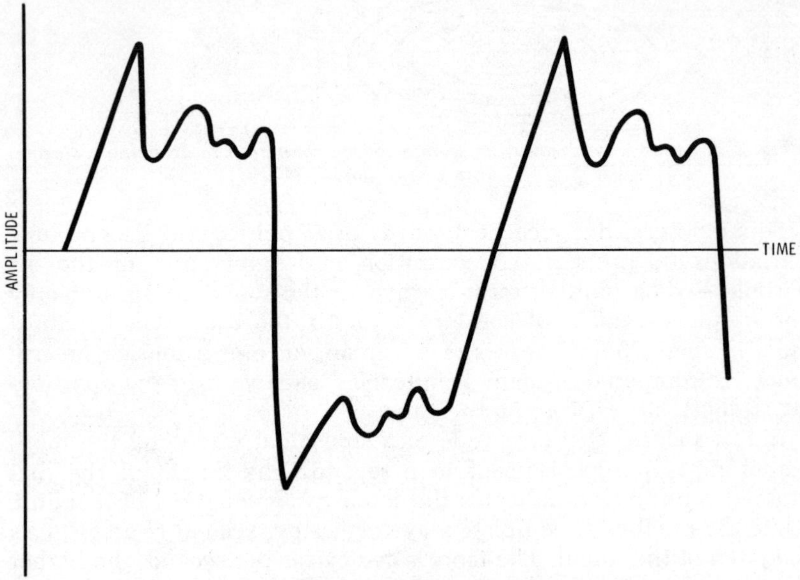

Fig. 2-4. A complex waveform.

The Decibel

In its resemblance to a variable amplifier, the human ear is ingeniously designed. It is sensitive to the faint sound of a scampering mouse and at the same time can filter the volcanic roar of a jet engine to a tolerable limit. Nature enables us to hear sound intensity logarithmically for our survival and protection. The louder the sound, the less sensitive our ear is to it. The variation in sound intensity between the dropping of a pin and the explosion of a bomb is many millions of decibels. If we were to suddenly hear linearly, or in exact proportion to the intensity of sound, it is doubtful that our hearing could survive the blast. Fig. 2-5 shows the difference between a linear and a logarithmic function.

Electrical energy is expressed and measured in several ways, such as in volts or watts. The volt is a unit of electrical potential and the watt is a measure of power. Because sound is received by a microphone and converted into an electrical wave in the recording process, the intensity of this wave is expressed in *decibels.* This is

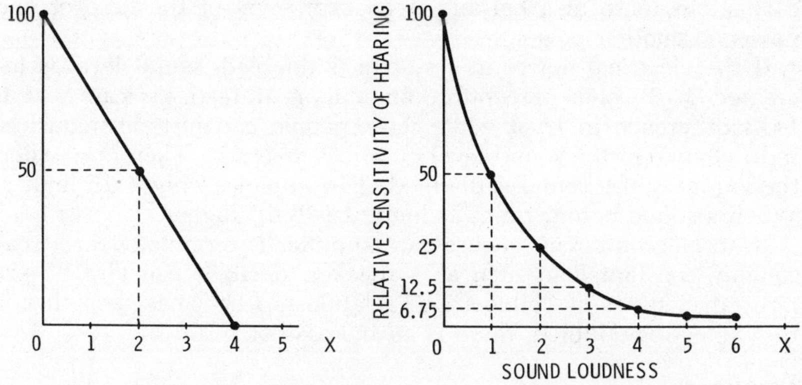

(*A*) *Linear.* (*B*) *Logarithmic.*

Fig. 2-5. Difference between linear and logarithmic functions.

proportionate to the log of a voltage or a power ratio, representing the acoustic properties of our hearing.

The *bel* is named after Alexander Graham Bell, pictured in Fig. 2-6, who discovered that humans hear sound intensity logarithmically. The audible range of sound is so large it is more convenient to use a smaller measure scale, *decibels*. The decibel (abbreviated

Fig. 2-6. Alexander Graham Bell.

dB) is one-tenth of a bel and is an expression of the ratio of one power to another power.

If the electrical power in a system is doubled, sound level is increased 3 dB, or a perceptible amount. A trained ear can hear a 1-dB difference in level while most people can just distinguish a 2-dB change in the sound level of voices or music. Therefore, when the loudness of a sound is doubled, it is, in effect, only 3 dB higher. When a sound is four times as loud, it is 6 dB higher.

If an electric circuit such as an amplifier is terminated in a reasonably constant load such as a speaker, decibels can also be expressed in terms of voltage. The relationship becomes such that if the voltage is doubled, there is an increase of 6 dB.

Frequency

The number of waves per second (the frequency of the wave) is measured in terms of hertz (Hz) in the case of both sound and electricity. A low-pitched tone causes the air to vibrate slowly, perhaps only 30 times a second, the lowest frequency audible. The highest frequency is generally between 14,000 and 18,000 Hz for young and middle-aged people. Older people often cannot hear above 5000 or 6000 Hz.

When the vibrating air waves reach the sensitive diaphragm in a microphone, they cause it to generate an alternating electrical current of the same frequency. As the diaphragm moves in, it creates a positive current, and as it moves out, it creates a negative current or vice versa.

Intensity

The intensity of the vibrations of the sound wave are also faithfully reproduced in terms of voltage, generally expressed in dB. The microphone translates a loud sound into a large voltage, while a softer sound results in a weaker signal.

The electrical waves of alternating voltage go from the microphone through several stages of voltage amplification; through a low-power stage; and then, transformed into current, they enter the recording head, which is actually an electromagnet. This process is clearly shown in Fig. 2-7.

ELECTROMAGNETS

Magnetic recording is an extremely complex process, involving a study of advanced physics. The mechanics of magnetic theory are complicated, and even experienced academicians sometimes become confused. The following explanation is intended to give a general idea of the magnetic-recording process rather than a spe-

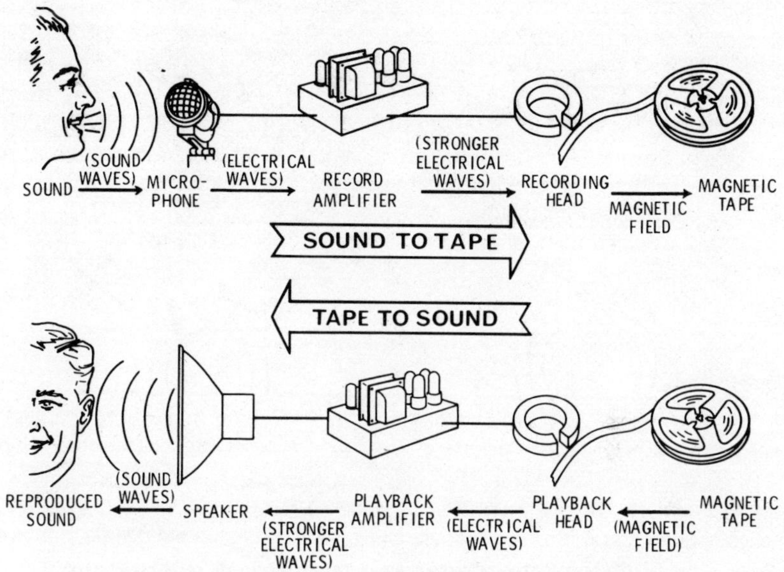

Fig. 2-7. From sound waves to tape and from tape to sound waves.

cific, highly technical explanation of how it works. No mention is made of either a direct-current or an alternating-current bias, since the theories involved are lengthy and require considerable use of higher mathematics and special treatment of the characteristics of magnetic materials.

An electromagnet is made by wrapping an iron core with wire to form a coil. When current flows through the wire, the core becomes a magnet. One end, or pole, of the iron core is a north pole. The other end is a south pole. If the direction of the current flow changes, the polarity of the electromagnet is reversed.

Around any magnet is an area of magnetic attraction commonly known as a *field,* which is illustrated in Fig. 2-8 in terms of lines of magnetic force. The closer the lines are spaced, the stronger is the magnetic attraction. These lines of magnetic force also have direction. Physicists have arbitrarily defined force lines as going from north to south outside the magnet, then completing their circuit by going from south to north inside the magnet.

A strong current produces a strong magnetic field, while a weaker current results in a correspondingly weaker field. Similarly, an alternating-current cycle causes each pole of the electromagnet to change in polarity through one complete cycle from north to south and back again to north. Both half cycles of the alternating current are shown in Fig. 2-8C.

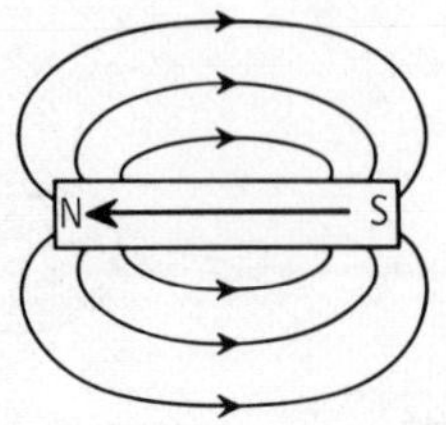

(*A*) *Permanent bar magnet.*

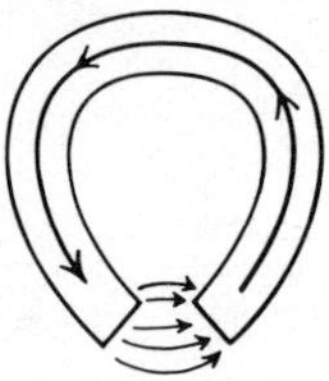

(*B*) *Permanent horseshoe magnet.*

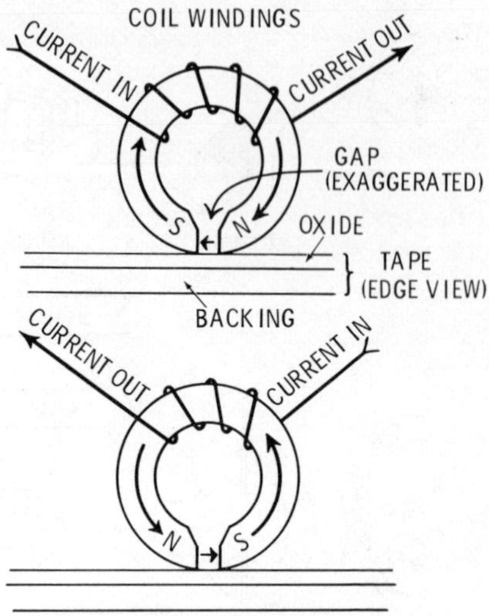

(*C*) *Electromagnetic* (*nonpermanent*) *recording head.*

Fig. 2-8. Similarity between permanent magnets and magnetic recording heads.

The Magnetic Recording Head

In the case of the recording head shown in Fig. 2-8C, the core is cut in a rectilinear shape with the poles as little as 0.00004″, or 0.04 mil apart. This distance is referred to as the *gap* of the head.

The cores of the recording head are magnetically "soft." That is, the head becomes magnetized easily and instantly when current flows into the coil, but loses its magnetism just as rapidly when the current stops. The recording head is unlike a permanent magnet or even the oxide on the tape itself, which is made from a magnetically "hard" material. The iron-oxide coating on the tape, being magnetically "hard," will hold any magnetism induced in it for an indefinite period of time.

Whenever a surge of positive current from the microphone goes through the coil, it magnetizes the recording head in one direction. When the current alternates and sends a surge of negative current into the coil, the head is magnetized in the opposite direction. Thus, the polarity at each pole will alternate between north and south according to the direction of the current.

TAPE MAGNETIZATION

When the iron-oxide–coated tape is in contact with the recording head, it offers an easier path than the air gap for the magnetic lines of force to follow. Therefore, following the course of least resistance, most of the magnetism gets across the gap by flowing through the iron-oxide–coated tape.

Although the magnetically "soft" iron ring of the electromagnet loses its magnetism when the current stops, the magnetically "hard" coating on the tape retains its magnetism, and the magnetized area becomes a small bar magnet itself (Fig. 2-9). Because the lines of force left inside the tape point in one direction, that direction must necessarily be north. The other end, therefore, becomes south. This is shown at point A in Fig. 2-10.

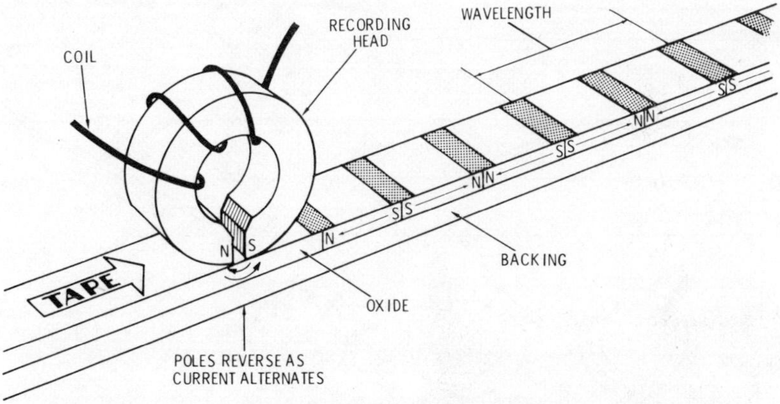

Fig. 2-9. Magnetized tape becomes a series of small bar magnets.

At point B, the current entering the coil is zero at its point of alternation and, consequently, does not create a magnetic field at that time. As a result, the tape moves a fraction of an inch without being magnetized any further. At point C, when a surge of negative current comes into the coil, a magnetic field in the opposite direction is set up, causing the polarity of the electromagnet to reverse. Again, the lines of magnetic force at the poles find it easier to flow through the iron-oxide–coated tape than across the air gap. Thus the tape is permanently magnetized, but this time in the opposite direction.

At point D, the tape has again moved, but since the current is not flowing, no new lines of force are set up at that point. As a result, the surges of alternating current leave the tape permanently magnetized by setting up a series of force lines of opposite polarity, creating a series of bar magnets on the tape.

Because the tape is moving, these poles occur at recurring intervals along the tape in a definite pattern. The frequency of the signal and the speed at which the tape moves determine the distance between poles, while the strength of the current, or voltage, determines the magnetic strength of each pole.

As shown in Fig. 2-10, the magnetic pattern on the tape for a 30-Hz tone consists of 30 magnetic fields pointing toward south, alternating with 30 magnetic fields pointing toward north. In effect, the oxide coating of the tape is broken up into 60 individual bar magnets, laid end to end every second the tape moves across the gap. On a tape recorder operating at 7.5 ips, the 60 bar-magnet patterns woud cover a space of 7.5 inches on the tape. A 10,000-Hz note would be represented on the tape as 20,000 such magnetic patterns in the same space.

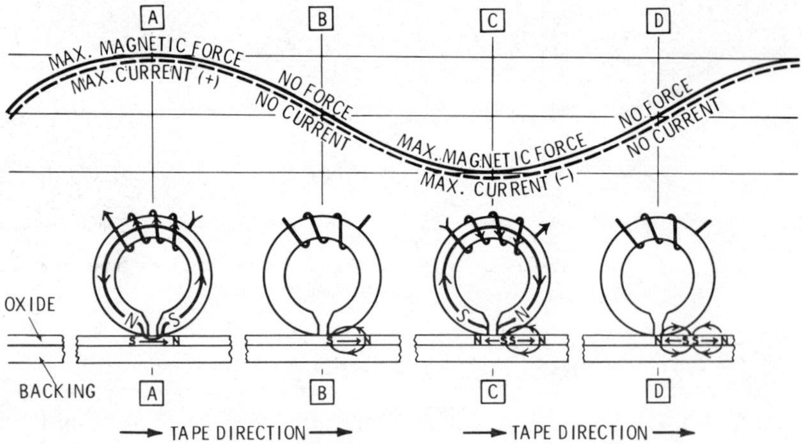

Fig. 2-10. Magnetization or recording process.

Some of the actual magnetic patterns on tape can be made visible by a simple process. A short piece of recorded tape is dipped into a solution of lighter fluid and carbonyl iron, then is allowed to dry. As the lighter fluid evaporates, the very fine particles of carbonyl iron will remain magnetically attracted to the tape in definite patterns visible to the naked eye (Fig. 2-11). A much more definite pattern can be observed by viewing the tape through a microscope as shown in the photograph of Fig. 2-12.

The heavily magnetized intervals on the tape (the poles) attract the carbonyl-iron particles, which appear as narrow lines across the tape. The stronger the pole, the heavier the line. The lower the frequency of the current, the greater the distance between lines. The actual wavelength of the original tone is equal to twice the distance between the lines.

Fig. 2-11. Using a solution of lighter fluid and carbonyl iron on tape to make the tracks visible to the naked eye.

The Recording Process

In summary, sound waves pulsate through the air and cause the diaphragm in the microphone to vibrate accordingly. This produces corresponding electrical pulsations in the microphone which are boosted in strength by an amplifier. The amplified pulsations are then fed into the recording head where corresponding magnetic fields are created, which, in turn, leave their magnetic patterns on the tape.

The Playback Process

In playing back a recorded tape, the recording process is somewhat repeated, only in reverse. During the recording process, an electrical current in a coil was used to create a magnetic field. In playback, a magnetic field moved through a coil is used to create an electrical voltage.

A basic principle of electricity is that a surge of electrical voltage can be generated by moving a bar magnet (or its surrounding magnetic field) through a coil. By moving a series of magnets (or their fields) of opposite polarity through a coil, an alternating electrical voltage will be produced. In the recording process, the tape was figuratively broken up into just such a series of short, permanent bar magnets.

Fig. 2-12. Microscopic enlargement of visible magnetic tracks.

The voltage is actually produced by the change in the magnetic field from positive to negative. The peak surge of current comes at the moment the polarity of the field is changing most rapidly. Therefore, when the magnetic force in the head is at its maximum, no voltage is produced. See point A in Fig. 2-13. At the point where the polarity of the head is reversing, the maximum voltage is produced as shown at point B in Fig. 2-13.

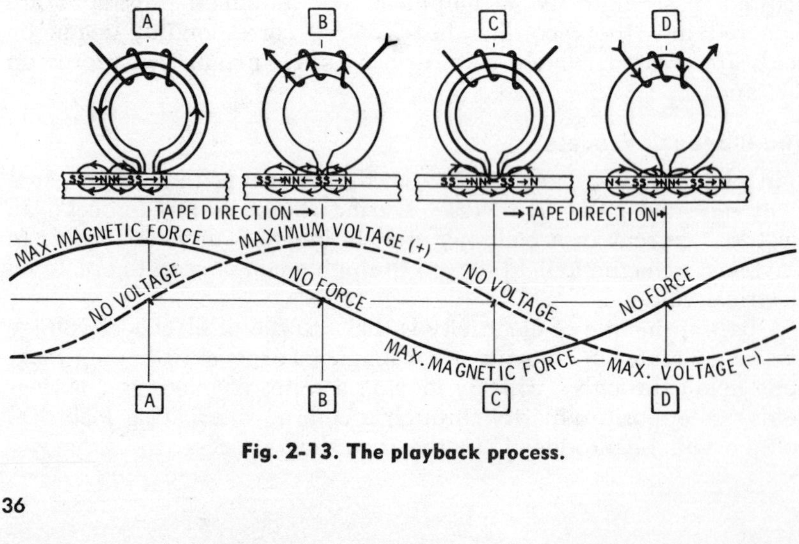

Fig. 2-13. The playback process.

In the playback process, the bar magnets on the tape are not actually moved through the coil of the electromagnet; however, part of the magnetic field of each bar magnet is. What happens is that the iron ring of the electromagnetic playback head temporarily routes the bar-magnet field through the coil.

In Fig. 2-13 at point A, a north pole and a south pole are on either side of the gap. Normally, the lines of magnetic force stay close to the tape, but because it is easier for the magnetic field to follow the iron ring (a much better conductor) than to jump the air spaces at the gap, it does exactly that. At point A, therefore, the magnetic force in the head is at its maximum, but the voltage is zero.

When the tape has moved a fraction of an inch farther (as at point B), a strongly magnetized line—a south pole this time—is at the gap. The iron ring of the electromagnet serves no useful purpose to the field, so it is ignored, and magnetic strength in the head is reduced to zero. However, since this is the point where polarity of the head changes most rapidly, maximum voltage is produced.

At point C, the situation again occurs where one pole is on one side of the gap and an opposite pole on the other. As at point A, the magnetic field again takes the easiest route and flows through the soft iron ring. Maximum magnetic strength is produced in the head but no voltage, since this is the point of alternation in the voltage wave from positive to negative.

At point D, the polarity of the head is again at the point of reversal, and consequently this sudden change in magnetic force results in the maximum surge of voltage. Since the surges of voltage alternate between positive and negative with the same frequency as that which was recorded on the tape, they can be amplified and fed through an amplifier to a speaker to reproduce the original sounds.

CHAPTER 3

The Tape Transport Mechanism

The tape transport mechanism is the term given to the parts of the tape recorder responsible for controlling the movement and handling of the tape. These parts operate in a basically simple manner. Most people with mechanical ability can study the mechanism and determine how it works and what each part is supposed to do. In many respects, tape recorder mechanisms are less complicated than record changer assemblies since there are no change cycle or trip operations. However, the requirements made on the tape mechanism are more stringent.

A record player rotates a prerecorded grooved disc on a turntable while a moving stylus transcribes the signals from the disc. A tape recorder pulls a tape past a stationary transcribing device, a magnetic head. Both machines require a recording medium and a transcribing device designed to operate at a constant speed. However, pulling a tape past a magnetic head at a constant speed is definitely a more complex operation than rotating a disc on a turntable.

A record player turns a single shaft at one, or at the most, four forward speeds. In contrast, the tape recorder capstan (shaft) must be able to operate at several speeds in both a backward and a forward direction. Besides normal recording and playing speeds, the tape recorder must have a fast forward speed and a fast backward speed (rewind speed). The fast forward and rewind speeds allow the rapid location of any portion of the recording. A complicated system of clutches and brakes is needed to facilitate rapid changes in tape direction and speed, yet maintain sufficient tension to avoid tape slack.

While the capstan is responsible for controlling the movement of the tape, the take-up and supply reels are responsible for its storage and handling. These reels are available in three types of containers: reel-to-reel, cartridge, and cassette. Each requires a transport mechanism carefully designed especially for it and able to handle the tape speedily and correctly at all times.

Fig. 3-1 shows a typical reel-to-reel deck. Note the straight line path of the tape from left to right. When the recorder is switched into either record or play mode, the heads and drive mechanism engage the tape. The tape is pulled from the full reel, fed past the playback or record head at a constant speed, and then wound onto the empty reel.

Courtesy Radio Shack

Fig. 3-1. Stereo recorder using basic reel-to-reel arrangement.

Tape movement is controlled by a single-lever control. The control can be switched to any one of five positions. These are: REWIND (fast reverse), STOP, RECORD, PAUSE (which allows the user to control the tape for short movements), and FAST FORWARD. To record, the RECORD button and function switch must be operated simultaneously. The slide lever immediately above the level window is a speed control that selects one of three available speeds, in this case

1⅞, 3¾ or 7½ inches per second (ips). Additional controls provide for adjusting the level on RECORD and PLAYBACK and for recording or playing back in single channel or stereo. Some recorders include a mode that simultaneously plays one channel while recording another.

Although the reel-to-reel is not as simple to load as cartridge or cassette recorders, it has a versatility the other two do not possess. Simplicity is sacrificed for the finer details of flexibility, such as choice of stereo or single channel, selection of more than one speed, option to mix multiple inputs, and so forth.

Cartridge transports are even less complicated in many respects than the reel-to-reel or cassette. The tape moves only in one direction and at one speed; there is no fast movement to complicate the control mechanism. The ends of the tape are spliced together, forming an endless spool of tape that never leaves the cartridge. Insertion usually automatically engages the starting mechanism.

An extra feature of the cartridge recorder may be a track selector which enables the user to choose any track, or pair of tracks in the case of stereo. Some cartridge players have an automatic track changer triggered by a sensor. The logical place to change tracks is at the join of the two ends. For this reason, the splice is marked with a dab of conduction surface material. When a sensor contacts this material, it activates a solenoid and cam mechanism which change the head position. In some machines, an additional solenoid automatically ejects the cartridge when the final track has been played.

Fig. 3-2 shows the drive mechanism of a cartridge model, which consists of a single belt from the motor to the capstan flywheel. On this particular machine, a two-speed drive allows the user to locate a spot on the tape faster than he could at the standard playing speed.

Fig. 3-3 shows a top view of the same recorder. At the left is the etched circuit board for the amplifiers. To the right is the transport mechanism; the motor housing can be seen in the foreground. Just to the right and above the motor housing, a small etched board carries the channel-selector mechanism. A rotating cam moves the head assembly across the tape and also turns on the indicator lamp by means of the switch on the etched board. The head assembly is just above the etched board.

At the top right side of the picture is the slot where the cartridge is inserted or "plugged." The cartridge engages and is held until the latch is released by a solenoid, which may be operated by an eject switch or by the contact on the tape when it is playing the last track.

The simplicity of the cartridge, from the transport mechanism viewpoint, is that there is no take-up or rewind mechanism to be driven by the motor. The tape is handled only by the capstan. Pressure against the capstan is provided by a pressure roller in the cartridge itself.

Rapidly gaining in popularity is the cassette. It is the easiest kind of recorder to use for straight forward recording and playback. The user never handles the tape as it is contained inside the cassette on two miniature reels and never leaves the tape container.

The cassette transport operates in much the same manner as the reel-to-reel. Like the reel-to-reel, it requires an additional drive besides the capstan, which is responsible for maintaining constant speed past the head(s). Fig. 3-4 shows the underside of a cassette-type recorder. The motor pulley with the heaviest belt is toward the upper right of the picture. The capstan pulley and flywheel are below it. The other pulleys, belts and rollers that can be seen engage and disengage by movements coupled to the controls, usually the push-button type. As these interactions and interlockings are fairly complicated, the details will not be explained here.

Fig. 3-5 is the underside of a drive mechanism that provides separate capstans for moving the tape in both directions (only one being engaged by its pressure roller at a given time). The motor is reversible, and fast movement in either direction is achieved by identical changes in drive and release from capstan contact. Fig. 3-6 shows the top view of the same mechanism.

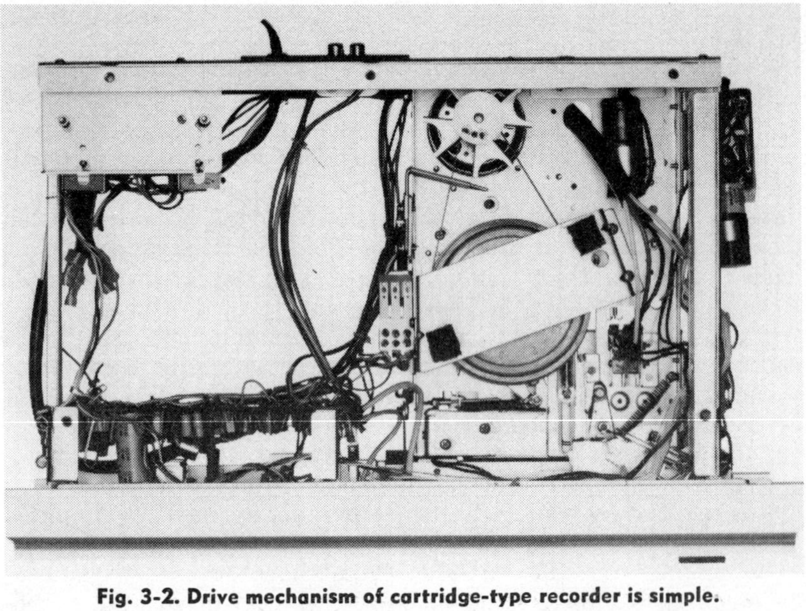

Fig. 3-2. Drive mechanism of cartridge-type recorder is simple.

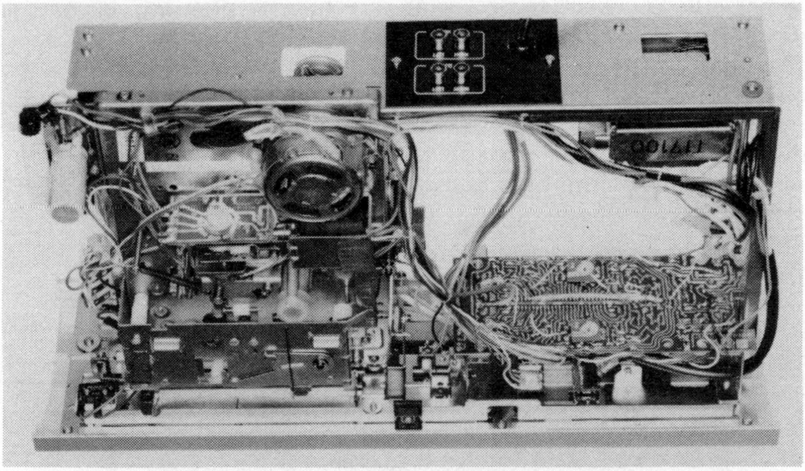

Fig. 3-3. Top side of recorder shown in Fig. 3-2.

The cassette has a unique feature that the reel-to-reel does not have, an interlock mechanism that prevents accidental erasure of a tape. The removal of tiny knockout blanks in a cassette of blank tape triggers a sensor in the mechanism of the machine which prevents the use of the record button and thus accidental erasure. Once

Fig. 3-4. Underside of cassette recorder showing drive mechanism.

the cassette knockout for relevant direction of play has been removed, it is impossible to do anything but play back the tape.

Another advantage of a cassette is removal of the tape at any point, either when recording or playing, without ruining the tape or losing the place. The cassette can then be reinserted later on to finish recording. Some cassette machines have a mechanism that senses the end of the tape and stops the machine or gives a tone signal to alert the user.

Fig. 3-5. Underside of typical reel-to-reel recorder showing drive mechanism.

One of the chief reasons for machine designers' headaches is tape spillage or breakage. Tape is especially vulnerable to this when it is stopped during a high speed operation. Probably the least expensive and most widely used solution is a push-button control system that sequences the action, particularly the stop button. All mechanisms are coordinated to act quickly yet smoothly. The drive is stopped by cutting off power to the motor and the brakes are applied. However, if the user presses the stop button indecisively or slowly, power may be cut off before the brakes are applied. The tape may spill off the reel or otherwise escape from its assigned path. Press the controls firmly and quickly to avoid this.

Some more sophisticated machines have eliminated dependence on the human operator. The moment the electrical contact is made, the action takes place, whether the button was pushed slowly or quickly, gently or firmly.

Fig. 3-6. Top side of recorder shown in Fig. 3-5. Note reel drive mechanism, and row of heads.

Coordination of controls, tape rewinding and threading, and operation of the recorder at a fast forward speed are all secondary functions of the transport mechanism. Its primary function is to move the tape at a uniform speed during record and playback. The other functions are necessary to the preservation of the tape, but ultimate success is only achieved from proper movement of the tape.

SPEED VARIATIONS

Maintaining a constant speed is critical in all tape recorders and has required a major amount of engineering time to satisfactorily solve the problem. Simplified controls have added immeasurably to tape-recorder popularity and have helped mushroom the magnetic-recording industry's sales growth since 1947. But the problem of simplifying the operation of the tape recorder is easy contrasted to devising a positive method of obtaining constant tape speed.

Long-Time Speed Variations

There are two types of speed variations affecting the performance of a tape recorder. They are: (1) long-time speed variations and (2) instantaneous speed variations.

Long-time speed variation is especially important in radio and tv broadcast work. If a radio station makes a 29-minute recording, it is imperative that it last precisely 29 minutes, not 30 or 28 minutes.

There are several reasons why tape does not run at a constant long-time speed: (1) Tape is pulled by a capstan and pressure roller, which permit a certain amount of slippage to take place. (2) Tape itself expands and contracts with changes in temperature and humidity, causing the playing time to vary. (3) Tape-recording mechanisms often use nonsynchronous- or induction-type motors in which speed is affected by variations in load and line voltage. Even on machines using synchronous motors, the capstan flywheel assembly is generally driven by a rubber puck or belts which are not positive in their drive (Fig. 3-7).

In tape recorders the tape is pulled by a capstan, a shaft around which the tape is wrapped. Therefore, the tape is not positively driven. In a positive-drive system, teeth in one gear engage teeth in another, or teeth on a sprocket engage perforations in a film, providing a drive in which slippage is impossible. In a friction drive, one roller turns another roller, both rollers having a smooth surface. Traction is obtained by friction and is not positive. In any type of friction-drive system, a certain amount of slippage is always present. Gear drives introduce flutter which requires an elaborate filtering mechanism to remove. Excessive cost of positive-drive systems and perforated film for these machines is the prime reason for using friction drive and tape for most recording applications. However, positive drive is necessary in motion-picture recording to synchronize the sound with the lips of the actors.

Expensive tape recorders designed for critical broadcast work generally have speed variations on only a few seconds in a 30-min-

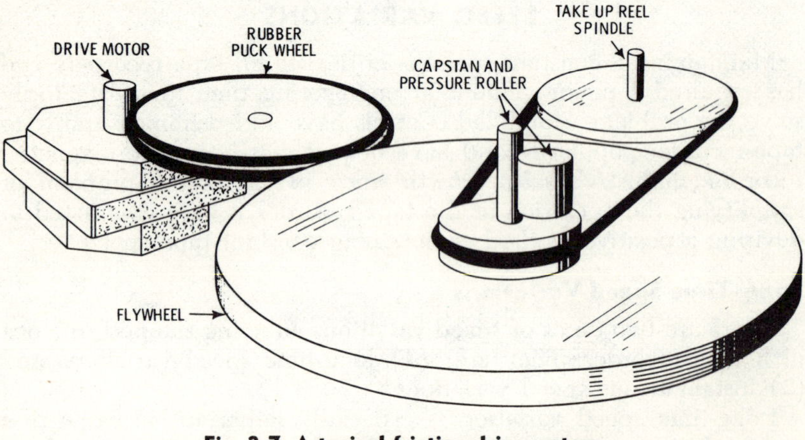

Fig. 3-7. A typical friction-drive system.

ute recording. In the less expensive home machines, the speed variation in extreme cases might run as much as several minutes in a 30-minute recording.

Apart from the necessity of maintaining exact speed in professional broadcasting and recording work, long-time speed variations also affect true pitch. Although no wow or flutter is present, if a record is running slow, the pitch will be low. If music is played at other than true pitch, the effect is disconcerting. Even though the variation is slight, musicians and music lovers with a trained ear are able to detect the pitch change.

Instantaneous Speed Variations

Another critical feature of a tape drive mechanism is the importance of instantaneous speed stability. Instantaneous speed variations are of an exceedingly short duration, generally less than a fraction of a second or, at the most, a half second. Speed variations of a short duration cause wow and flutter, twin menaces to good recording results. The wow meter, shown in Fig. 3-8, will measure these variations in speed which are then expressed in percentages.

Fig. 3-8. A wow meter used to measure tape speed variations and express them in percentages.

Among the factors causing instantaneous speed variations are these: (1) Flat spots which develop on the rubber rollers. (2) Slipping of the drive rollers and tape on the capstan. (3) Sticking of the tape to the heads or guides. (4) Bent or eccentric shafts. (5) Motor "hunting" or "clogging." (6) Irregular feed or take-up reel operation.

Wow and flutter are particularly noticeable even to untrained ears. They are without question the most objectionable features of low-priced tape recorders. It is interesting to note that the ear can detect instantaneous speed variations of as little as 0.1%. Many of the earlier home-type tape recorders had instantaneous speed variations above 0.5%.

Wow is a relatively low-frequency speed variation, noticed especially on sustained notes such as in piano or organ music and chimes. When wow is present, a musical note that is held will sound unsteady. Its pitch will waver, producing a very irritating effect.

A still shorter speed variation will result in flutter. Flutter causes the sound to be garbled, again showing up in piano music especially. Often when a recording does not sound clear, the cause will be flutter, even though it may not be diagnosed as such. High-frequency flutter is often mistaken for other types of distortion.

CAPSTAN DRIVE

The capstan drive is the very heart of a tape recorder. The tape is pulled by a roller pressing the tape against a rotating shaft. The rotating shaft is known as the capstan. In many older machines rubber or composition covered shafts were employed to pull the tape. The rubber capstan was attached to a common shaft with a flywheel and was motor driven. The early Brush, Eicor, and International Electronic Company tape recorders, for example, used a rubber roller to pull the tape. The early Eicor tape recorder is shown in Fig. 3-9.

However, the introduction of lubricated magnetic tape in 1949 necessitated a more positive drive and required the use of a pressure roller against a steel capstan (Fig. 3-10).

Importance of a Large Capstan Diameter

In order to have the necessary stability, a fairly large diameter capstan is required. If the capstan is at all eccentric, it will pull the tape faster at one point in its revolution than at another, introducing serious wow and flutter. A small capstan will also have a tendency to whip. To obtain the required rigidity, it is generally necessary to use large diameter capstans.

However, large capstans have one main disadvantage. If the capstan size is increased, the flywheel size must also be large, since more inertia is needed to stabilize a large slow-speed capstan than one that is small but high speed. The stabilizing effect of a rotating mass is proportional to the mass and to the square of its speed of rotation, as well as to the square of the diameter at which it is distributed.

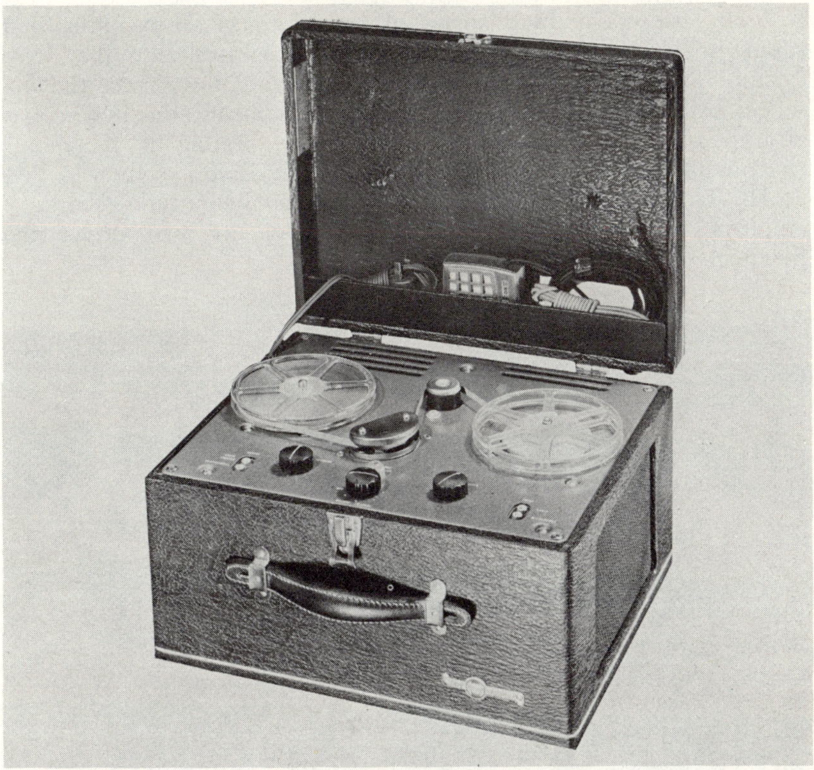

Fig. 3-9. An early Eicor tape recorder.

When tape speeds of 15 ips were common, larger capstans were practical. But in the small recorders that use 1⅞ ips, a larger capstan would have to rotate so slowly that it would have little stabilizing effect. Small capstans are more practical in these recorders so the shaft can rotate at a speed sufficient to achieve some stability.

Using flywheel mass to achieve stability has proved impractical in modern recorders. Instead, efforts have been concentrated on the design of precision motors that achieve higher stability in driving torque so that stabilization is less dependent on the flywheel mass. The capstan size of most recorders varies from slightly under an eighth of an inch to approximately five-eighths of an inch.

Methods of Motor Drive

In order to stabilize the capstan rotation and provide the most uniform tape speed possible, whatever flywheel is used is attached to the capstan shaft. The motor then drives the flywheel by either a puck-, belt-, or direct-drive system.

Puck Drive—The high speed of the motor shaft requires that some method of speed reduction be incorporated into the drive system. The puck drive is a method of driving the rim of the flywheel by a friction process. In some mechanisms, the flywheel is equipped with a rubber tire and is driven directly by the motor shaft as in Fig. 3-11A. Another type of puck drive is shown in Fig. 3-11B. This method, which is used by many home-type recorders, employs an intermediate-puck roller which in turn drives the flywheel.

Fig. 3-10. Close-up of head assembly with guide (center) and drive capstans (either side of heads).

The rim-drive method, although lacking some of the possibilities of the other drive methods, results in very satisfactory performance, even when produced to meet competitive prices. For rim-drive applications, the motor shaft is usually equipped with a small-diameter metal pulley. This pulley is a bushing positioned on the shaft by means of a set screw.

For the intermediate-puck drive system a rubber-tired wheel, called the idler wheel (see Fig. 3-12), is mounted so that the motor pulley will drive it. The idler wheel, in turn, will drive the capstan

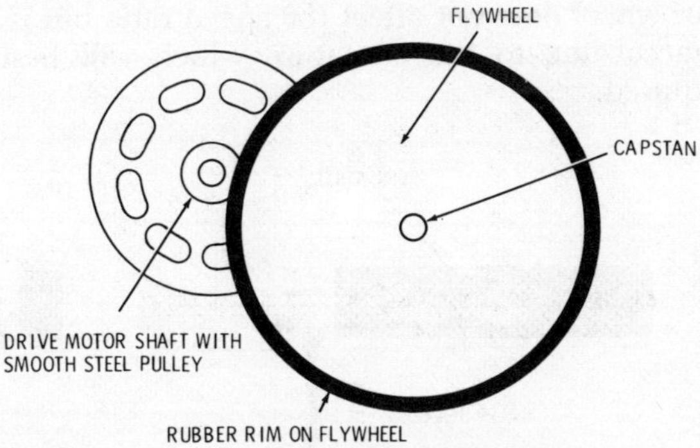

(*A*) *Rim drive on flywheel.*

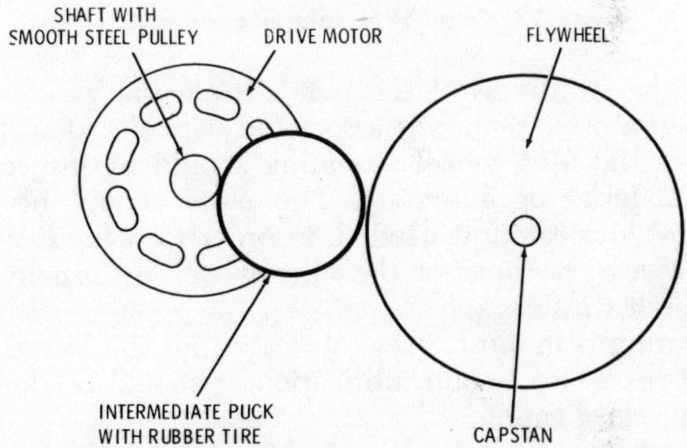

(*B*) *Intermediate puck drive.*

Fig. 3-11. Two methods of capstan drive.

flywheel rim. The center bearing of the idler wheel is mounted on a plate. Although there are design variations in the mounting of this bearing, the purpose of each is threefold: (1) Ensure a correct wedge angle of the idler wheel between the flywheel rim and the motor pulley, exerting enough traction to transmit the torque and yet not stall. (2) Keep the idler wheel movable horizontally so that spring tension may be applied to the bearing mounting in the direction of the wedge angle, providing constant pressure of the idler wheel against both the flywheel rim and the drive pulley. (3) Maintain alignment of the idler wheel in a vertical plane with the motor pulley and the flywheel rim. An exploded view of a typical bearing mounting system is shown in Fig. 3-13.

The diameter of the motor pulley is determined by the diameter of the flywheel, the speed of the motor shaft, the capstan diameter, and the loss in working the rubber of the drive tire. The diameter of the idler wheel does not affect the speed ratio but is selected and designed according to the curvature which will best deliver the torque required.

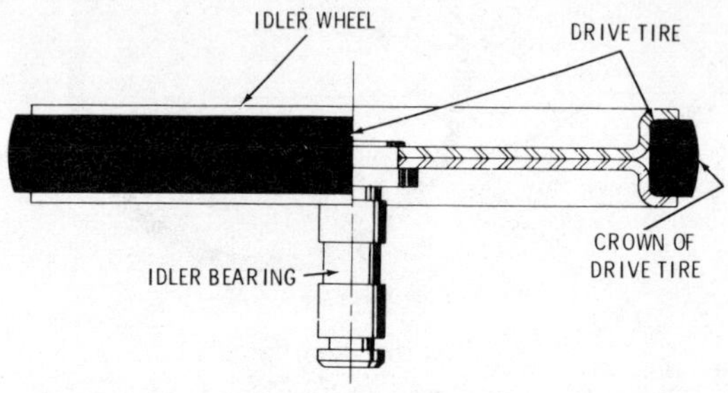

Fig. 3-12. Detail of an intermediate puck wheel.

Due to the sturdiness of the motor itself, the greatest source of trouble in rim-drive motors is associated with the idler drive wheel. All parts of the idler wheel mounting should be inspected for extraneous material or gumming. The parts should be thoroughly cleaned and lubricated, if needed, in order to prevent speed reduction. The sleeve bearings of the idler wheel are usually lubricated through a felt washer.

This bearing is in most cases lubricated for the life of the motor; if it is felt necessary to add lubrication, it should be done carefully to avoid overlubrication.

The tire of the idler wheel is of natural or synthetic rubber, and after being cemented to the idler wheel, it is ground with a crown to be absolutely concentric with the axis of the idler stud. The radius of the crown and the hardness of the rubber are design factors dependent on the torque required. For this reason, if a damaged tire is encountered, it is advisable to replace the entire idler wheel with an exact duplicate.

Puck-drive systems give very good instantaneous speed regulation, tending to reduce wow and flutter. However, the condition of the surface of the rubber rollers is of utmost importance. Indiscriminate lubrication or an "oil bath" treatment are very bad practices. Oil has a tendency to creep or run over surfaces, spraying the rotating puck with a minute amount of oil. This results in slippage and deterioration of the rubber parts. One of the most common troubles encountered is slippage due to a film of oil on the idler tire.

If slippage results from excessive oil, it is best to replace the idler wheel and pucks after first cleaning all nearby surfaces. These include the flywheel rim, motor pulley, and idler plate. Use a cloth moistened with trichloraethylene. If the idler wheel is not replaced,

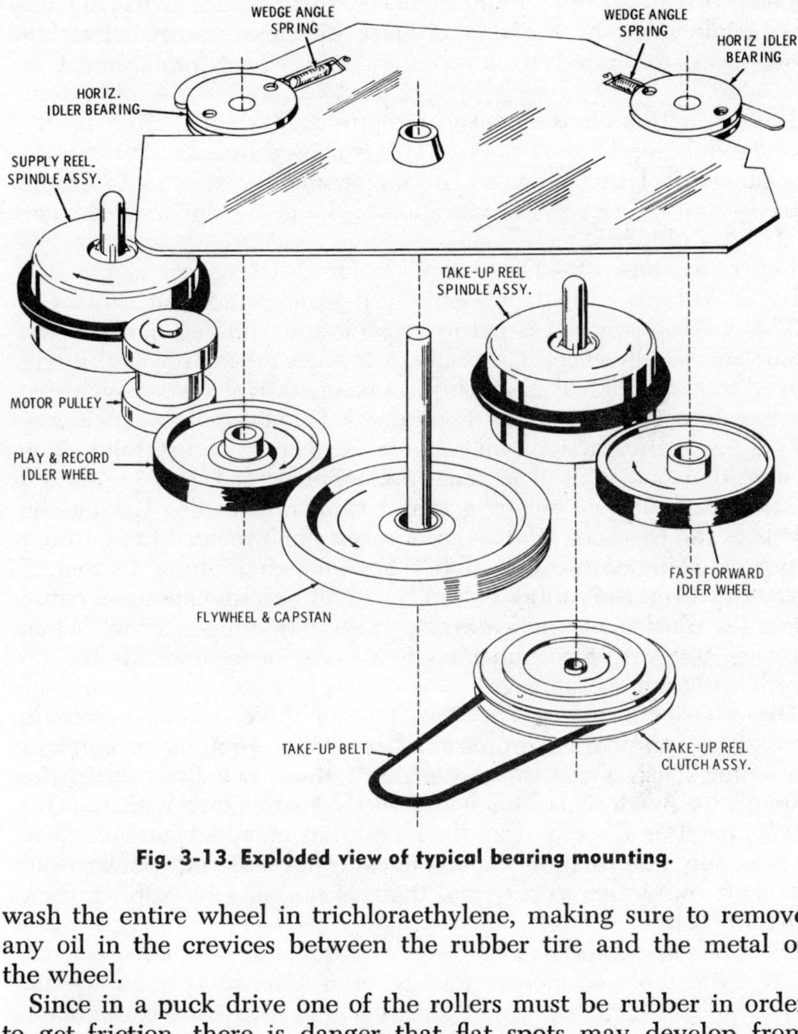

Fig. 3-13. Exploded view of typical bearing mounting.

wash the entire wheel in trichloraethylene, making sure to remove any oil in the crevices between the rubber tire and the metal of the wheel.

Since in a puck drive one of the rollers must be rubber in order to get friction, there is danger that flat spots may develop from leaving the roller engaged with one of the shafts, thereby indenting the rubber. When a rubber roller develops flat spots, it makes a disagreeable thumping noise and introduces wow each time the flat spot makes contact during rotation. This dent can sometimes be removed or minimized by running the recorder for a half hour before using.

Time and wear will develop a glaze on the surface of the rubber, resulting in slippage. The glaze may be removed by carefully holding a piece of fine sandpaper against the tire while the idler wheel is being driven by the motor. However, this practice is usually not worthwhile since by the time a glaze develops, the material has probably deteriorated to a point where wheel replacement is advisable.

Flutter, a tremulous quality of reproduced sound, often results from modulation by motor noise. This is caused by : (1) Insufficient mechanical isolation between the motor and the remainder of the drive system. (2) Excessive vibration of the motor due to imbalance of its rotating parts.

Belt-Drive Systems—The second major drive system is the belt drive. Two types of belts are employed—composition and rubber.

The composition belt is flat in construction, running between the motor and the flywheel. Generally, a tension roller arm is also employed to keep the belt taut. This is a comparatively expensive drive because an extra idler roller is required. However, it is fundamentally a good drive system, making for excellent speed stability. The composition belt is used in some professional recorders.

The other type of belt is a round rubber belt, also linking the motor to the flywheel. The use of a rubber belt means that an idler is not needed because the rubber has enough tautness to remain tight. However, the rubber belt in time will become loose and cause slippage. Hard spots may develop, also introducing wow. When slippage develops, the rubber tends to wear away and fills the recorder with rubber shavings.

Direct-Drive Systems—The third type of drive system, becoming increasingly popular in professional recording equipment, employs the motor shaft as the capstan shaft. If there is a flywheel in the system, the flywheel is attached directly to the motor shaft. This drive, operating directly from the motor, has many advantages since there is only one rotation shaft with no transmission of power from one shaft to another. Therefore, there is no need for rubber puck rollers or belts.

To rotate the capstan and have a reasonable capstan diameter, relatively slow speed motors must be used. For example, to operate at 7.5 ips and have a capstan diameter of approximately .24 inch, one needs a 600-rpm motor, which is expensive, large, and heavy. Professional equipment can tolerate a motor of this type; however, its cost, size, and weight usually preclude its use in portable home-type equipment.

Still another disadvantage of the direct drive is that irregularities in motor core structures and variations in damping as the motor armature rotates tend to introduce a certain amount of wow and flutter in the tape. These irregularities, however, have been successfully minimized by careful design of the motor and by special production techniques and quality control.

Motor Types

Any drive system operating from an ac supply can use two types of motors, induction or synchronous. A conventional induction motor usually has a fairly stable speed characteristic. It has good instantaneous speed regulation, including very little flutter and wow. It is inexpensive, compact, well cooled, and produces a lot of power for its size. Nonsynchronous or induction motors, however, are subject to speed variations that are caused by changes in load and line voltage.

The other type of motor in use is a synchronous motor. The synchronous motor has absolutely perfect long-time speed regulation since its speed is controlled by ac power alternations. However, it tends to introduce a certain amount of instantaneous speed variation during its rotation, resulting in some flutter. It is an expensive motor, runs hot, and its size is disproportionately large for its amount of power.

While ac motors provide reliable drive, the modern trend, especially since the advent of transistors, has been toward battery operation. In recent years, high-reliability, low-consumption motors with extremely good constant-speed characteristics have been developed for use with batteries. Precision tolerances during manufacture have improved their capability.

Another possibility, used by some manufacturers, is electronic speed control which generates a synchronous ac source from the batteries. This provides extreme precision in speed control. As evidence of the economy possible, the Nagra minirecorder, shown from two views with the covers removed in Fig. 3-14, can operate on two penlight batteries for 7½ hours. With extra thin tape, 0.015 inch wide (the same width as used in cassettes), and playing at $^{15}/_{16}$ ips, the recorder will play or record for the same length of time with only one loading of tape. It handles musical programs "adequately," although understandably it does not have the "superquality" achieved by larger machines with the same standards.

TAKE-UP REEL DRIVE

The capstan, of course, is used to pull the tape at a constant speed past the magnetic head. However, tape moving at a constant speed must then be spooled onto a take-up reel. Because the tape goes from a small diameter at the beginning of the reel to a large diameter at the end, the reel speed must change since the tape speed remains constant. The rpm of the reel changes considerably from a large to a small diameter. Therefore, all recorders employ some type of variable-speed drive.

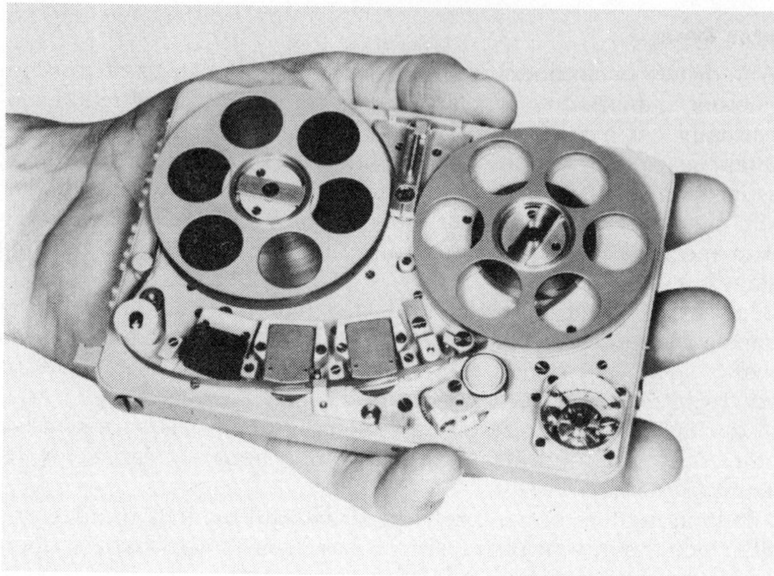

(A) Top view.

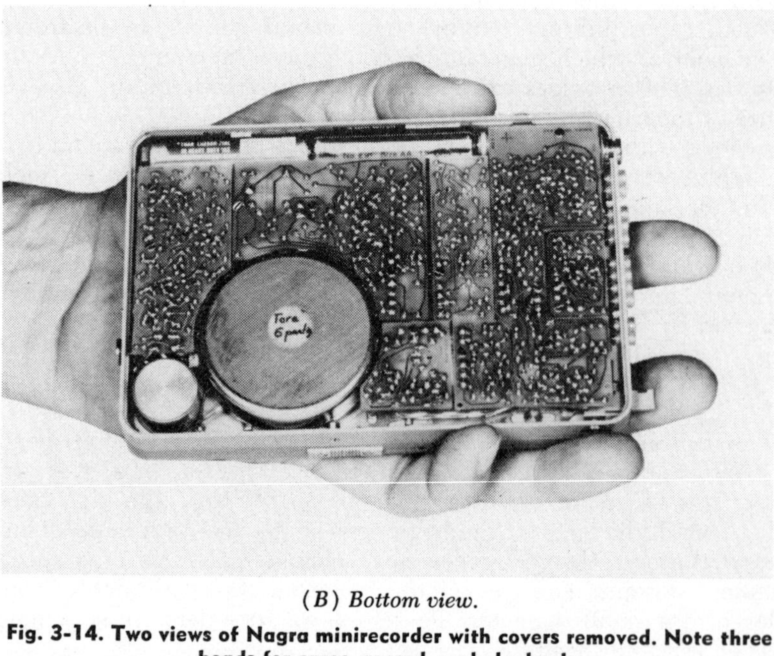

(B) Bottom view.

Fig. 3-14. Two views of Nagra minirecorder with covers removed. Note three heads for erase, record, and playback.

The slipping clutch is a common variable-speed drive. Generally, the clutch consists of a rotating disc that is driven at a constant speed, often by the same motor that drives the capstan. Next to this disc is placed a piece of felt. The felt slips on the disc and drives another disc which, in turn, drives the reel shaft.

The slippage between the drive disc, the felt, and the reel disc applies a more or less constant torque on the take-up reel. This constant torque is something of a disadvantage, since when the reel is nearly empty, the tape radius is short and tension on the tape will be high. However, when the reel is nearly full, tension on the tape will be low since the tape radius is long.

Because the tape is being pulled past the head by a capstan, in theory the tape should be moving at a constant speed. However, due to the change in reel diameter, the tape is under heavy tension at one time and under light tension at the other. The tape tends to slip by the capstan at a slightly faster rate under heavier tension. Because of this variation, it is desirable to have constant tension on the take-up reel. The slipping clutch device used on most tape recorders does not provide constant tension. To help solve this problem, tape manufacturers have developed thinner tape, permitting an increase in the hub diameter of the reel, thus reducing tension difference between the inner and outer reel diameter as illustrated in Fig. 3-15.

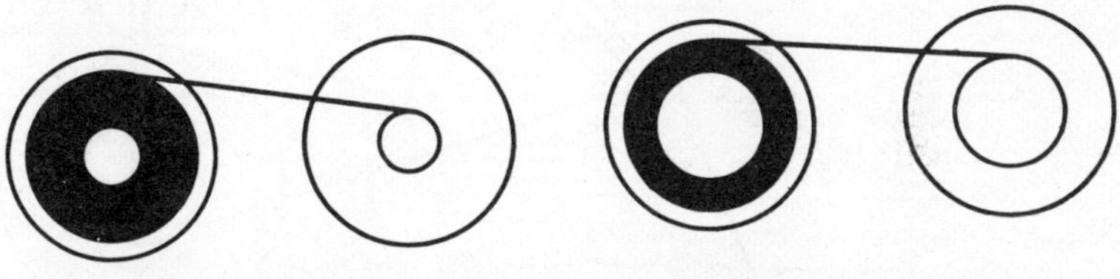

Fig. 3-15. Using thinner tape permits increased hub diameter.

The other method of revolving the take-up reel is by use of a high-slip induction motor. This method is used by several of the professional recorders. It is also used on some home recorders. In this method of variable-speed drive, the reel is attached to the motor shaft or is driven by the motor shaft. The motor is not allowed to run at its full speed. Instead, it exerts a torque on the reel. By careful design of the motor, a nearly constant tension can be maintained regardless of how full the take-up reel may be. Also, the motor take-up is more dependable since its torque is unaffected by changes in temperature, humidity, and lubrication as is the slipping felt-clutch device.

A third method of driving the take-up reel is by a slipping cloth belt on a polished pulley. The belt is driven at a high rate of speed around a very smooth pulley which, in turn, drives the take-up reel. The belt drive suffers many of the same problems as the felt clutch since the tape tension will vary from time to time depending on lubrication, temperature, and humidity conditions.

Yet another form of take-up device can achieve close to constant take-up torque. The tape is passed around an idler pulley that adjusts its position according to tape tension. If the tension is too much, the variable resistor, or other device attached to the arm that carries the idler pulley, reduces the drive to the take-up motor so the tape slackens its tension. If the tension is too little, the resistor applies more power to the motor.

When the system is properly designed, the overall effect is that the idler pulley assumes a position according to the radius of the tape on the take-up reel, such that tension remains the same. Using a clockspring-type loading on the idler pulley arm ensures that the tension varies very little as the arm moves. All that the movement of the arm achieves is adjustment of the motor torque so that the tension is correct. Fig. 3-16 illustrates this system.

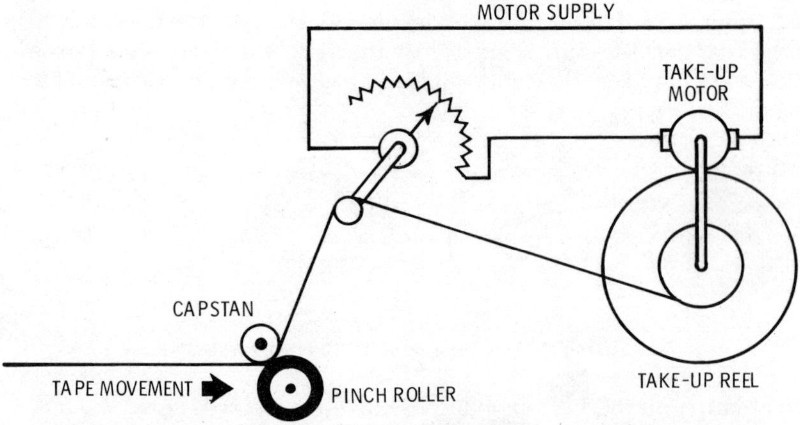

Fig. 3-16. System to allow take-up tension to be held within close limits.

REWIND AND FAST-FORWARD MECHANISMS

During both fast forward and rewind, little back tension is applied to the feeding reel. This allows the tape to be moved at a high speed with a minimum of mechanical power.

In home recorders a mechanical linkage generally shifts the reel shaft or puck shaft into a position to be driven by the motor. By using a minimum amount of mechanical power with little back tension, a "soft" tape wind is achieved. It is essential that tapes are softly wound to prevent physical distortion in which one edge of the tape is stretched.

If a tightly wound tape is subjected to a sharp change in humidity or temperature, tremendous tape tensions will be built up on the reel. Each layer of tape contributes a progressively stronger pressure on its adjoining inner layer which will often stretch the tape beyond its yield point, thus permanently deforming its shape. In cases of excessive physical distortion, the cumulative pressure becomes so great that the plastic reel hub shatters.

It is regrettable that the more expensive professional machines are the worst offenders in respect to tape distortion. These recorders, using a direct-coupled motor drive for either rewind or fast forward, invariably wind the tape "hard." The direct motor drive exerts a more or less constant tension on the tape during operation, thus building up a considerable pressure with each turn of the drive reel.

Many tape engineers feel that the constant-torque reel drive used on most home recorders contributes to the best tape storage, since the outer turns are wound with far less tension than the inner turns. However, as pointed out earlier in this chapter under capstan drive, the professional-recorder manufacturers are faced with the necessity of obtaining constant tension throughout the reel to eliminate capstan slippage. There is little doubt that the manufacturers of professional machines, faced with a choice between two evils, favor a direct-coupled motor drive with little capstan slippage over a constant-torque reel drive providing a soft tape wind.

TWO-SPEED DRIVE

In the professional field, two tape speeds are becoming standard throughout the industry. The 15-ips speed is being widely used for original recording where the highest possible fidelity is desired. However, with the tremendous improvement in performance at 7.5 ips, this speed has been widely accepted as standard for the bulk of broadcast work. Therefore, the professional machine should have both speeds.

In most professional equipment the speed change is now accomplished by means of switching the fields in the synchronous drive motor. Since most professional machines use this type of motor, it is possible to change from a four-pole motor, running at 1800 rpm, to an eight-pole motor, running at 900 rpm. A synchronous drive motor with six poles, running at 1200 rpm, can be changed to a twelve-pole motor running at 600 rpm. The speed change is simply accomplished by flipping a switch which connects the proper number of motor fields in the circuit.

By contrast, the speed of some makes of professional recorders can be changed by the physical transfer of capstans as shown in Fig. 3-17. A removable pressure roller and/or capstan of different diameter is used for different speeds. This method, while perhaps bothersome and time-consuming, serves the purpose. A well-machined shaft is used, making the removal and fitting of the capstan easy. However, dirt on the capstan shaft can make the capstan eccentric with resulting wow and flutter.

In home recorders, both the 7.5- and 3.75-ips speeds are important. The 7.5-ips speed is generally used for higher quality musical recordings, while the 3.75-ips speed serves primarily for voice applications. It is true that some recorders operating at 3.75 ips give better performance than others operating at 7.5 ips, but this depends on the design and precision of the recorder itself.

Most cartridge and cassette recorders are designed for one speed. Cassettes are standardized at 1⅞ ips; however, some are made for two speeds, the second of which may be 3¾ or $^{15}/_{16}$ ips.

It is more difficult to obtain good performance at slower tape speeds for two reasons: First, the high-frequency response is limited and the output is reduced. This subject will be fully treated in the chapter on magnetic heads. Second, the wow and flutter are increased since there is less inertia in tape systems operating at slower speeds. Also, any minute variation in tape speed has a tendency to adversely affect the recording when the tape is traveling at slow speeds.

In the home-recorder field, the majority of recorders change speed by changing the position of the intermediate puck from one diameter on the motor shaft to another, as illustrated in Fig. 3-18. Generally, the intermediate puck is moved between two steps of different diameters on the motor shaft.

While this system works very well as a speed-changing device, the method of shifting the pulley is of great importance. On some recorders the pulley is actually forced over the edge of the step. If the recorder is in a neutral or off position, the speed change can be easily and safely accomplished. However, if the recorder is running, the pulley will nick the intermediate puck roller. The nicked edge will result in serious wow and flutter.

The more desirable method of speed change used in many of the newer home recorders is accomplished by actually removing the intermediate puck from both the motor drive shaft and the flywheel. The intermediate puck is then shifted to its new position and is reengaged, eliminating any possibility of nicking its edge.

In most home recorders the intermediate puck roller is the heart of the drive system. The smoothness and consistency of the surface about its center axis are all important in keeping wow and flutter to

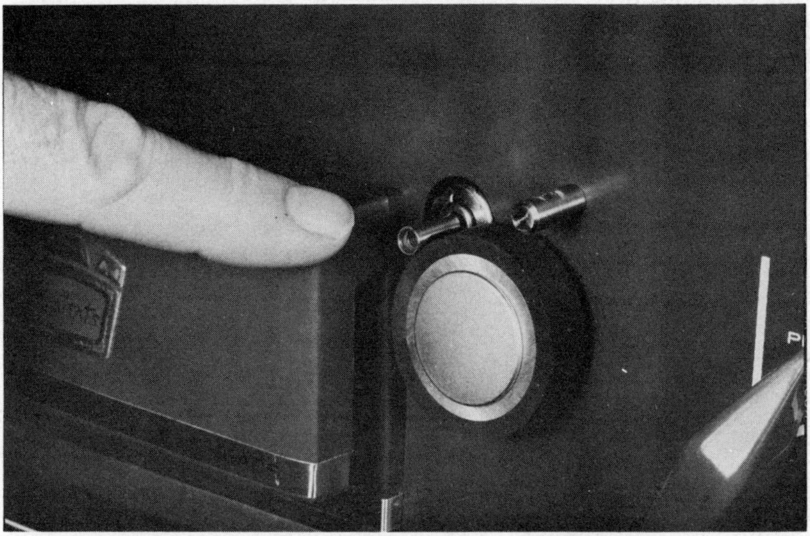

(A) *Capstan and pressure roller to be removed.*

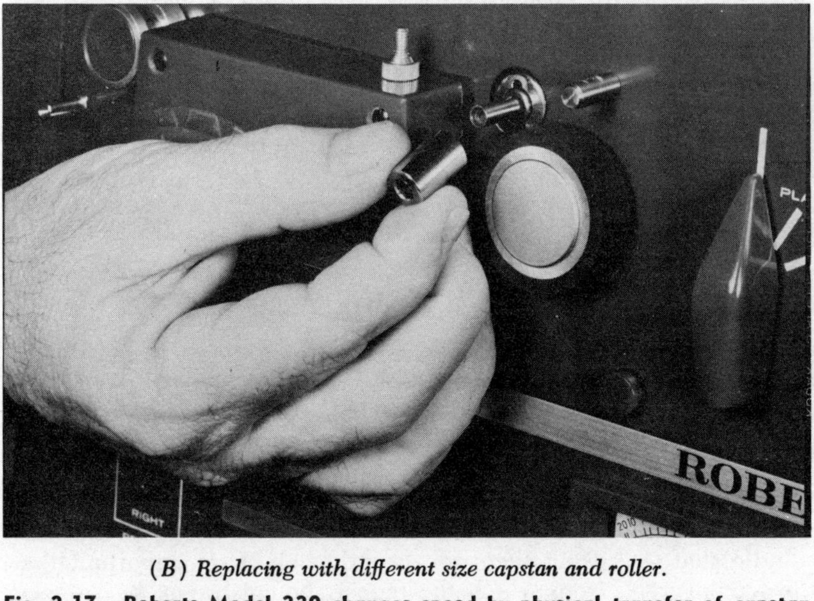

(B) *Replacing with different size capstan and roller.*

Fig. 3-17. Roberts Model 330 changes speed by physical transfer of capstan and pressure roller.

a minimum. Care must be exercised in operating those machines which do not remove the intermediate puck from the shaft, motor pulley, and flywheel when changing speeds. It is a wise precaution to check the types of speed change used by the recorder before attempting to change the speed.

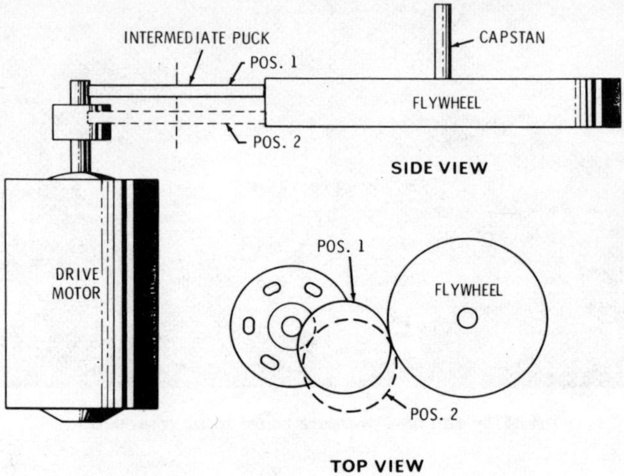

Fig. 3-18. Changing speed by changing puck position.

Several of the older home machines change speed by varying the capstan size. However, this method is gradually being discarded. Because of cost considerations in home machines, a tapered, precision-machined shaft cannot be used. Consequently, if the capstan fits well, it is difficult to remove and put on. If the capstan is easy to remove, the fit is so poor that the capstan will be eccentric, and wow and flutter will develop.

BRAKING MECHANISMS

When the recorder is put in the stop position, the reels must brake as rapidly as possible without causing damage to the tape. Generally, the brakes consist of a metal wheel which engages a rubber rim on the reel shafts. Also, a rubber or felt pad can be used which rubs on a metal disc on the reel shafts.

In normal operation the brakes are removed, but in the stop position they are suddenly released and applied to the revolving reel spindle shafts, causing them to abruptly halt. It is important that the feeding reel, from which the tape is being drawn, be stopped first. Otherwise tape spillage will result.

The adjustment of the brakes is critical in many machines to prevent tape spillage and still not cause physical damage or stretching of the tape. Several machines use no brake action whatsoever. The tape is braked only by reversing the direction of drive on the reel spindles. An interesting approach to braking is also used which applies braking pressure toward both reel spindles while stalling the motor, rapidly halting the tape with no danger of spilling or breaking the tape. Some professional machines use a form of braking produced by changing the motor connections from motor drive to a short-circuited generator. Such electrical braking applied to the reel drive spindles is extremely smooth and cannot snatch as mechanical brakes may.

POSITION INDICATORS

Position indicators are almost universally included. They are counters that are driven by the feed or the take-up reel and count the revolutions of the reel. This feature enables rapid location of selections within a reel of tape. Fig. 3-19 shows the position indicator on the Wollensak Model T-1400 recorder.

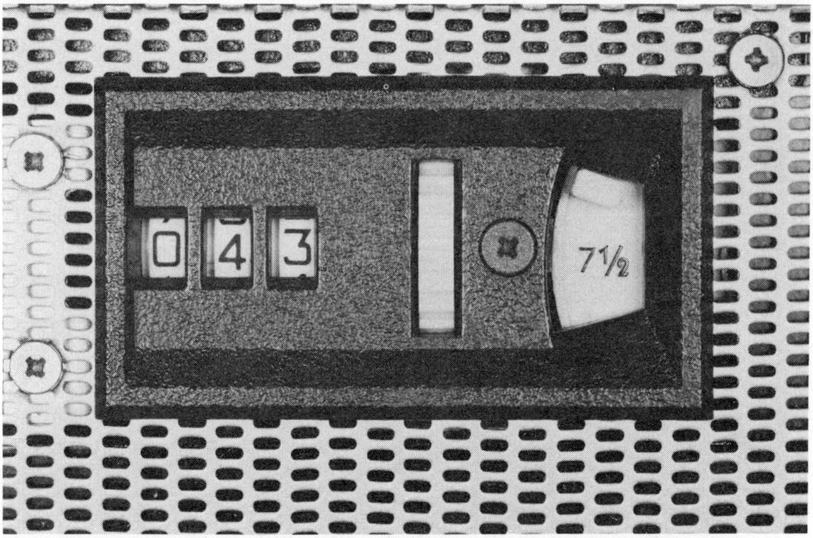

Fig. 3-19. Position indicator used on the Wollensak Model T-1400.

MAINTENANCE

Since a recorder motorboard is a mechanical device with shafts and bearings, occasionally a drop of oil is required to keep all parts exposed to friction or wear turning freely. Most recorders, however, require little oil. Because of the slippage problem caused by oil on the rubber rollers, excessive oiling is a real threat.

Most recorder bearings are manufactured from a powdered bronze material, such as *Oilite*. This is a dependable, life-long bearing in which oil is actually impregnated into the powdered bronze bearing material. Since most recorder shafts turn at slow speeds, it will last a long time.

Like any mechanism, the biggest menace to a recorder is dust, dirt, and grit particles working between the shafts and bearings, scouring them and causing wear. To keep any tape recorder at peak operating performance, the motorboard mechanism should be kept clean and free from dust and grit. When worn, an occasional shaft or bearing may require replacement.

CHAPTER 4

Tape Containers

In previous chapters, we have referred to the choices of tape containers available for recorders. The original form was reel-to-reel. From this mode two alternative "prepackaged" forms developed. The cassette, which uses ¼-inch tape, contains two small reels within a compact container easily inserted or removed from the recorder. The cartridge consists of a single reel or spool within a container which it never leaves.

In this chapter, we will discuss the requirements that each container has to meet in order to satisfactorily handle the tape for recording or playback processes.

REEL DESIGN AND CONSTRUCTION

The various sizes of reels for magnetic tape play an important role in the recording process. Reels for magnetic tape must meet a variety of exacting requirements. First, the reel must in no way distort the tape. Reels with large openings in their hubs in which the tape can become deformed are undesirable. As tape is wound on a reel, tremendous pressures are built up within the inner turns. Any threading slot, especially if it is large, will tend to deform the tape at that point.

In addition, the 7-inch plastic reel (the most common reel size) should have a 2¼-inch minimum hub diameter to avoid timing errors, as we have seen in the chapter on transport mechanisms. The timing error problem is perhaps the most serious consideration. Fig. 4-1 shows typical speed changes for a number of recording machines. It can be seen that recorders tend to speed up at the beginning of a reel due to high tension on the wind-up side and tend to slow down at the end due to high tension on the supply reel. Referring to Fig. 4-1, the percent of speed change for the 2¼-inch hub diameter (curve B) is much less than that of the 1¾-inch hub diameter (curve A). At the 2¼-inch hub diameter, most of the timing error occurring on the 1¾-inch hub reel will be corrected. A 2¾-inch hub gives still closer timing results and is used in applications demanding precision programming. A comparison of the three hub diameters used for 7-inch reels is shown in Fig. 4-2.

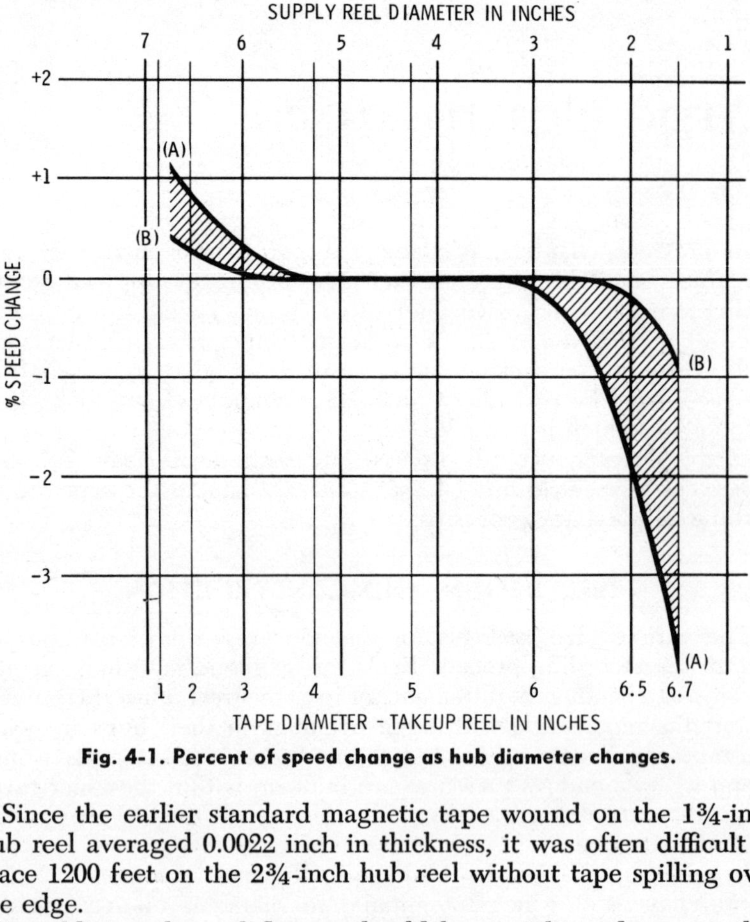

Fig. 4-1. Percent of speed change as hub diameter changes.

Since the earlier standard magnetic tape wound on the 1¾-inch hub reel averaged 0.0022 inch in thickness, it was often difficult to place 1200 feet on the 2¾-inch hub reel without tape spilling over the edge.

In addition, the reel flanges should be straight and true so that they will not rub either edge of the tape. There should also be a method of locking one end of the magnetic tape to the reel and securing it with several turns on the reel so that it will not pull free. The tape should lie flat against the thread-up device and not leave

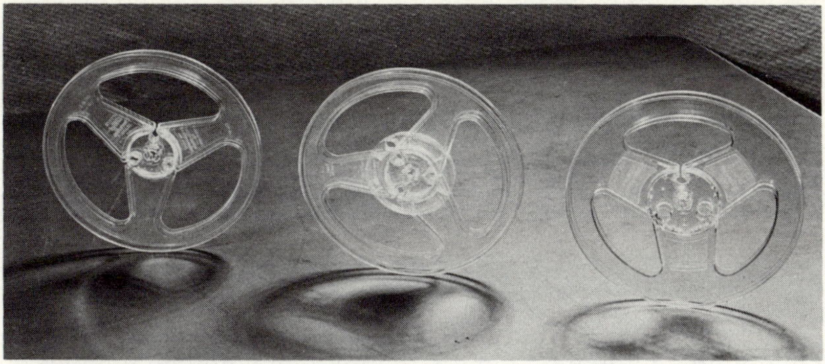

Fig. 4-2. Three types of 7-inch reels with different hub sizes.

a hump on subsequent turns. The NAB 10½″ plastic reel, for example, provides a unique post thread-up device.

On the large size NAB 10½-inch and 14-inch reels, the tape is supplied on either a metal hub or on a metal hub with flanges. Also, a plastic replacement for the 10½-inch metal hub with flanges with its large center hole is available. The major change is the reduction of distance between flanges to a value of .270 inch. This reel is manufactured from glass-filled plastic. With the changes in the distance between flanges, the winding characteristics have been greatly improved. A large center hole is used for those professional recorders that use a hold-down clamp to secure the reel to the pedestal. A supplemental 10½-inch plastic reel has also been developed. This reel has a small center hole, corresponding to the 7-inch reel, and contains other drive holes to adapt it to the recording machines of various manufacturers. This reel also contains both conventional and V-type thread-up slots as shown in Fig. 4-3.

Unless care in storage is exercised, this type of glass-filled plastic reel is subject to a certain amount of warpage, thus minimizing its improved winding advantages.

Because there is a cost saving in ordering 2400 or 4800 feet of tape on an NAB hub without flanges, some recording engineers prefer to screw on their own flanges as illustrated in Fig. 4-4. Also, several recorders operate using only a hub without flanges to eliminate any possibility of the edge of the tape scraping on the reel flanges, causing wow and flutter. However, on a properly designed tape recorder, the tape will ride between the flanges and at no time scrape on them.

A number of changes have improved the metal NAB hub. One improvement is grinding the surface of the metal hub. Since the hub is made from a die-cast mold, the surface of an unground hub is mottled, resulting in tape distortion near the center of the reel.

Also, grinding the metal hub makes the surface diameter exactly the same across the entire width of the hub. Because the hub has exactly the same diameter, the tape will not be distorted through receiving a long edge.

Fig. 4-3. The 10½-inch reel.

Another improvement is narrowing down the hub in 10½-inch hubs and reels. The hub was originally designed to be 350 mils thick. This left ample room between the tape edges and each flange. The purpose, of course, was to prevent scraping and rubbing on the flanges. However, during the fast forward and rewind on professional machines, the tape would wind in a sloppy manner, causing some turns of the tape to be higher than other turns. During storage when the flanges of the reel would be slightly compressed, the higher turns would be crushed. Consequently, there has been a recent trend to narrow the metal hub.

As previously mentioned, the plastic 10½-inch reel has automatically incorporated a narrowed-down hub in the plastic mold, requiring no further attention. The distance between flanges is reduced by molding a thicker plastic flange. While the 10½-inch plastic reel is subject to warpage, it is expected that with the solution of this problem the reels will be perfected.

(*A*) *Disassembled.*

(*B*) *Assembled.*

Fig. 4-4. The NAB 10½-inch reel.

Much has been done recently to increase the convenience as well as the fidelity of tape recording using reels. Eastman Kodak Company's *Thread-Easy Reel* (Fig. 4-5), for instance, is threaded simply by making one loop with the tape and drawing it through a slot in the reel. The reel also has a built-in splicing jig, index scale, and write-on panels.

A standard reel with a self-threading feature has also been introduced by the 3-M Company, which eliminates the "all thumbs" method of threading tape on the reel. The self-threading reel (Fig. 4-6) has solid flanges and when sealed with a plastic collar is a dust-proof, self-storing unit. This reel automatically winds the tape after it has been placed between the flanges near the hub.

CASSETTES

The early forerunner of the present cassette was the RCA "cartridge," so called before the words "cassette" and "cartridge" came

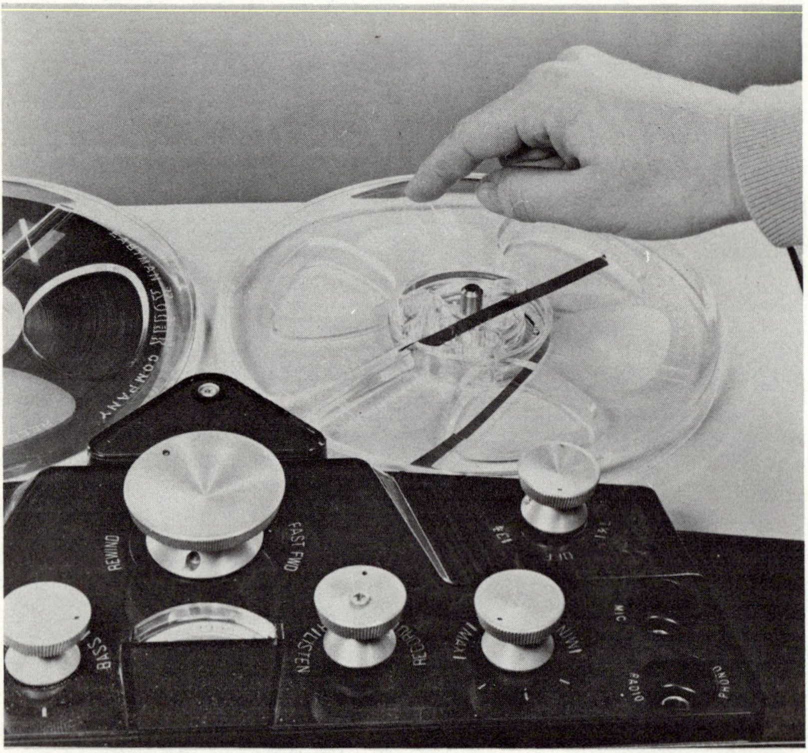

Courtesy Eastman Kodak Co.

Fig. 4-5. "Thread-easy" reel.

to have such well defined or distinctive meanings as they have today. Fig. 4-7 shows one of the RCA cartridges being inserted into a recorder designed to use it. It measured 7¼ inches by 5 inches by ½ inch thick and carried ¼-inch wide tape on spools whose hubs had a diameter of 1½ inches.

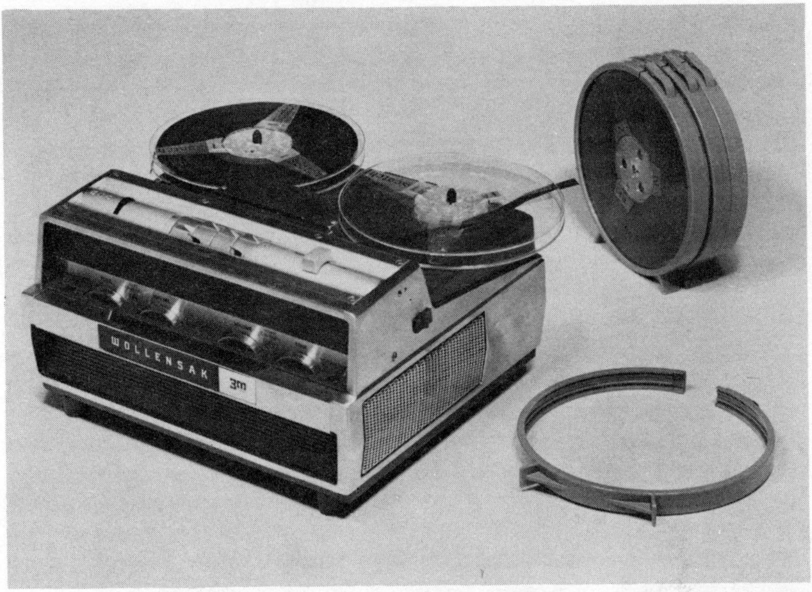

Courtesy 3M Company

Fig. 4-6. The self-threading reel.

The RCA cartridge had five openings along the front. The end two were for capstans that drove the tape in their respective opposite directions, and the middle three enabled the cartridge to be used with a 3-head machine, in whatever combination those heads were arranged. One way used separate heads for erase, record, and playback, like professional 3-head machines that allow immediate monitoring of the program just taped.

Another more likely arrangement of the three head apertures used those next to the capstan apertures for permanent playback heads and the center one for record. As machines of this type were likely to be used less frequently for record, this was designed in one of two ways. Either the record head could be stationary, in which case recording would always be done with the tape moving, say, from left to right, and the cartridge would be flipped to record the other "side." Or the record head could be movable, so that the complete cartridge could be recorded without removing it from the recorder. A choice of designs was an option with this type of cartridge.

The RCA cartridge also had slots that exposed the tape path between the capstan holes and the spools. Sensors on the machine detected when the end of the tape was reached by the tape pulling tightly at the feed-reel end. This either stopped the machine or reversed its direction of play, according to the way it was programmed.

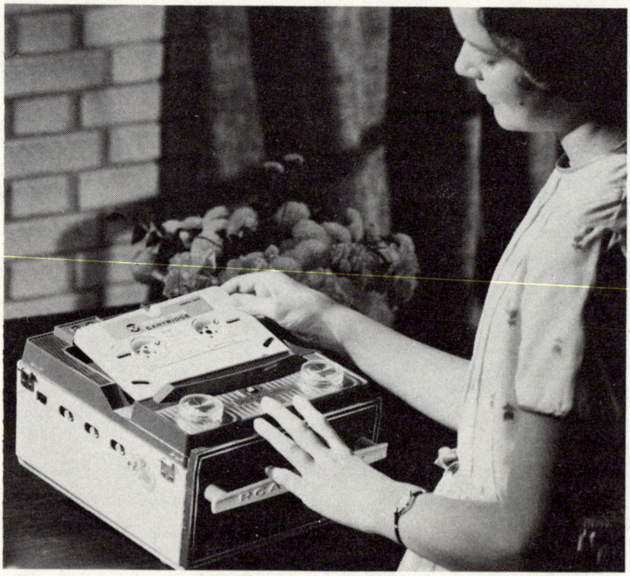

Fig. 4-7. The large size cartridge issued by RCA which is similar in many respects to the cassette in use today.

In the back of the RCA cartridge was a notch that exposed a metal insert that was spring loaded. When the cartridge was not in a machine, this metal insert served as a brake to prevent the tape spools from moving. When the cartridge was inserted into a machine, a projection in the compartment where the cartridge fitted depressed this metal insert against the spring inside the cartridge, releasing the brakes so the spools could function.

Newer cassettes do not have this feature. However, the carrying and storage cases designed for use with cassettes have some form of insert that engages the spigots inside the spools, locking the spigots in place and thus preventing the tape from unwinding and spilling. This serves the same purpose as the RCA device but in a less costly way.

Fig. 4-8 indentifies the features of the presently widely used cassette. Some other designs with interchangeable sizes have been used, but this is the one accepted as standard. In comparison with the size of the earlier RCA model, it is 4 inches by 2½ inches and is ½ inch thick only at the front where it engages the works of the machine. The main body is less than ⅜ inch thick ($^{11}/_{32}$ inch actual). It is usually stored in a box that allows for the full thickness at the widest place. Although it takes the same thickness as the RCA model, it occupies only a little more than one fourth the area.

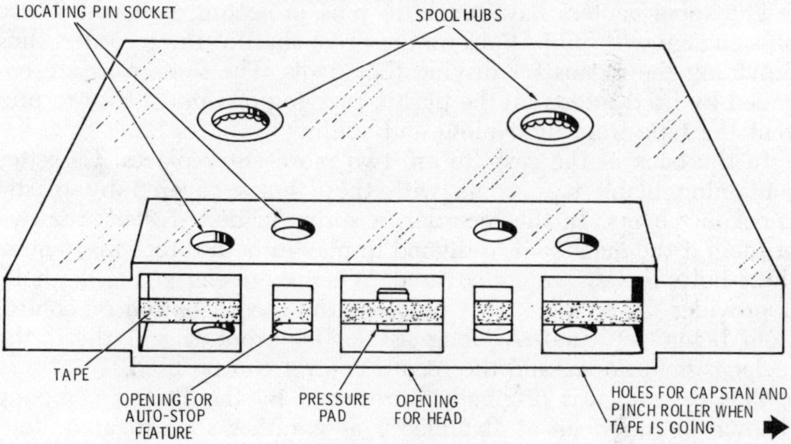

(A) *Front view.*

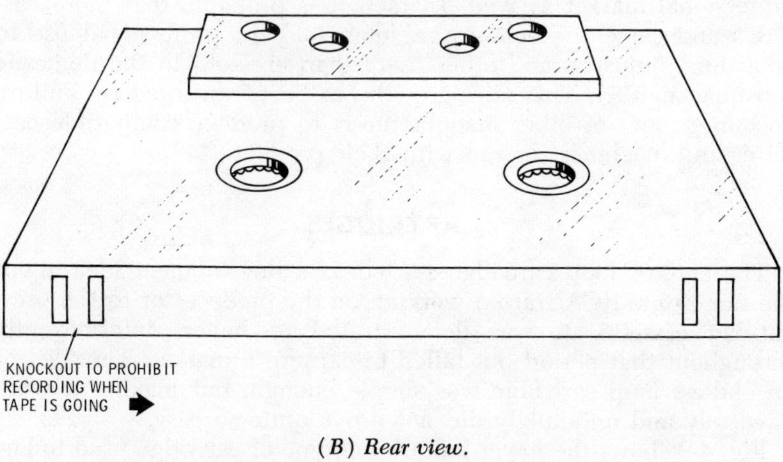

(B) *Rear view.*

Fig. 4-8. Details of the presently used standard cassette.

The holes for the capstan and pinch roller are similar to the RCA design. There is a single opening for the head in the center of the front with a built-in pressure pad behind the tape—a feature the RCA type did not have. The holes to enable auto-stop (or reverse) sensors to "feel" the tape are in the front between the larger holes.

Behind the tape, in the sides of the cassette at positions level with the holes in the front for the sensors to enter, are two more round holes, slightly smaller than those for the capstan. These holes are normally used to engage locating projections in the machine for precise positioning of the cassette.

The spool centers have six little pins projecting inwards. These pins engage with projections on the drive shaft of the recorder, thus providing the means for driving the spools. The same pins are engaged by fixed inserts on the plastic carrying or storage case to prevent the tape from unraveling and spilling.

In the back of the cassette are two more entry places. Cassettes containing blank tape come with these holes covered by plastic knockout covers. In the recorder, a spring-loaded "feeler" (or two of them if the machine is designed to play in both directions) enters these holes if they are not covered. When the feeler enters the hole, it provides an interlock that prevents the record button or control from being activated. As long as the knockout is still there, the feeler is held down, and the record control can be used.

This cassette was originally introduced by the Philips company (known in American as "Norelco") as a rather sophisticated "toy" for the domestic consumer market. However, coupled with its compactness and the advent of higher quality tapes that came along about the same time, it has become a very useful alternative for the professional market as well. In fact, it is probable that more cassettes and cassette machines are now sold for professional use to educators, industry, and other users than are sold to the domestic consumer market. This sales growth has been expedited by Philips' encouragement of other manufacturers to produce compatible cassettes and machines for use with their product.

CARTRIDGES

The endless loop cartridge started as a development idea about the same time RCA started working on the predecessor to the present-day cassette. It was demonstrated to buyers intermittently throughout that period but failed to capture a market. The idea of an endless loop cartridge was simple enough, but making it work flawlessly and unfailingly did not prove quite so easy.

Fig. 4-9 shows the lower half of one type of cartridge used today. The white plastic spool is free to rotate on its spindle. The tape is

Fig. 4-9. Lower half of one type of cartridge.

pulled from the inside by the capstan and the pinch roller. After the tape has passed the capstan it is fed back onto the outside of the spool.

Here is where operation is critical and requires careful design modifications as well as careful attention to loading the tape in regular production. If the tape runs too slack, it may back up in the space outside the spool compartment behind the pinch roller until it gets caught with the tape going around the post to the pressure pads (lower left). Compartment barriers should prevent this, but if the tape "piles up" anywhere but on the spool, it will probably jam.

This is an argument for having no more tape than will conveniently wind onto the spool. The tape should not fit tightly, but neither should it be loose. On the other hand, if pulling the tape from the center causes the tape being fed back onto the outside to tighten, the tape will jam where it pulls out from the middle. Either way the tape jams.

Tape handling must be carefully controlled. Even a slight buckling of the tape fed to the outside of the spool will increase the diameter of that outside turn. The spool will start taking up the slack faster than it is drawing tape from the middle. As this continues, more of the spool will have a larger turn size than originally, until again the tape jams.

The appeal of this type of cartridge is that no rewinding is needed. Some cartridge-type recorders provide for "fast forward," which moves the tape at a speed faster than normal playing speed in order to advance it to a particular spot faster than at normal playing speed.

Because of the problems with this cartridge in its early development, another type was developed by CBS Laboratories and marketed in conjunction with the 3M Company. It used the narrower tape of the present cassette at a playing speed of 1⅞ ips, so a lot of program could be squeezed onto a small spool of tape. Its container was about 3 inches square. Fig. 4-10 shows such a cartridge.

The end of the tape was accessible outside the cartridge and was picked up by the machine and fed through the playing mechanism onto a take-up reel in the machine. When the tape was completely played, the machine automatically rewound the tape back into its cartridge. The machine was also equipped with a sequencing device that enabled it to play a stack of tapes, rather like the way a record changer handles disc records.

Fig. 4-10. A self-loading cartridge developed before the standard cartridge was adopted.

With the cartridge used today (Fig. 4-9), the only side of the tape accessible to the machine is the coated side. There is access in three places—at the two rectangular openings backed by the pressure pads and at the opening for the capstan, where it is backed by the pinch roller. By using one opening for the head and the other for electrical sensing contacts, a dab of conducting material applied to the juncture of the tape ends can activate a sensor which will trigger the track change or eject mechanism. The eject mechanism can be either mechanical, solenoid operated, or both. Only by having it solenoid operated can the machine change tracks or eject the cartridge when the playing of a track is completed.

AUTOMATIC TAPE CHANGERS

The advent of cartridges and cassettes has made feasible the design of changers that function much like those used for discs. In fact, tape changers can store far longer program selections than a disc changer ever could. We have already mentioned the changer built for the cartridge of Fig. 4-10. Cassettes and the newer regular type of cartridges also are built with automatic changers.

Courtesy Telex Communications Division

Fig. 4-11. Automatic cartridge changer.

Fig. 4-11 shows an automatic stereo cartridge changer. Here twelve 8-track cartridges are arranged in a circle. The player mechanism can play successive tracks on one cartridge or one track in turn from each of the dozen cartridges, depending on the mode selected. It can also be programmed for selection like a juke box.

CHAPTER 5

Drive Motors

The drive motor is one of the most critical parts of the drive mechanism since it must supply the power which drives the tape. The main requirement of any tape-recorder motor is that it hold a constant, normal-running speed. The quality of the recording and subsequent playback is, of course, dependent on the linear velocity of the tape which is, in turn, dependent on the rotational speed of the capstan. Any deviation in the capstan rotation speed will affect tone or pitch. The audiophile with the legendary "golden ear" will easily recognize that the tape speed is not perfect, even though the variation may be exceedingly minor.

Another requirement for motors is that they must resist overheating despite being small and lightweight. In addition, the suitability of the tape-recorder motor is determined by such factors as elimination of stray magnetic fields, ruggedness, dependability, silent operation, and infrequent maintenance.

Three types of motors are generally used in tape-recorder design: the ac induction motor, the ac synchronous motor, and the dc battery-operated motor. Each of the three types will be discussed in this chapter.

INDUCTION MOTORS

The vast majority of small home-recorders use a single-phase shaded-pole induction motor. The shaded-pole induction motor provides an acceptably constant running speed for home recorders and is small, compact, simple in design, and very inexpensive. With normal use the motor proves rugged and practically indestructible. A typical induction motor is shown in Fig. 5-1.

The induction motor is able to withstand relatively long periods of "stalled" operation without serious overheating. This condition may occur when some part of the drive mechanism jams and the whole system, including the motor, is blocked until it receives attention.

Courtesy Arvin Industries Inc.

Fig. 5-1. A typical shaded-pole type drive motor.

The motor speed, which is commonly stepped down to the capstan speed by rim driving, is slightly under 1800 rpm in the majority of recorders. The average power consumption is around 60 watts.

Induction Motor Components

An induction motor is composed of a stationary field structure (the stator) and a rotating element (the rotor), as shown in Fig. 5-2. The stator consists of coils wound on projecting iron cores or poles. The open end of each core, the pole face, extends toward the rotor. The pole faces partially encircle the rotor.

The rotor is mounted in the center of the fields. This rotating element is a cylindrical laminated iron core carrying lengthwise conductors in slots on its surface. The conductors are shorted by a ring at each end of the core.

The purpose of the stator is to provide while remaining stationary a revolving magnetic field around the rotor. The stator magnet is placed so that its pole is near the rotor conductors, and the magnet is effectively revolved around the rotor. As the magnetic field revolves, it causes the lines of force to cut the rotor conductors.

How the Induction Motor Works

The induction motor works with a relative slippage between the rotating magnetic field, set up in its stator or field coil, and the rotation of the armature or rotor.

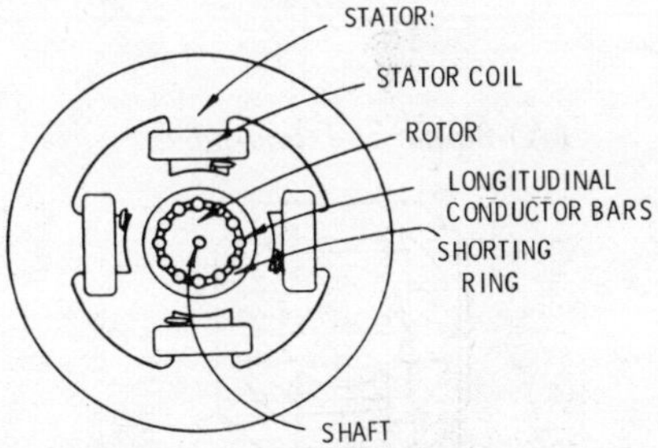

Fig. 5-2. Construction of an induction motor.

As we have seen, the induction motor consists of two main parts, the stator and the rotor. The stator, the stationary part of the motor, is fastened to the frame. The rotor or armature, the rotating member of the motor, is supported by bearings on each end and supplies the power to the recorder.

The stator is equipped with magnetic poles around which are wound the field coils. These poles are electromagnets and, of course, are supplied by alternating current going on and off at the rate of the line frequency, or the poles are going through a complete cycle, north and south poles, 60 times a second (Fig. 5-3).

The number of poles determines the speed of the motor. It is easy to see that in a two-pole motor the magnetic field reverses twice every cycle, and cycles occur 3600 times a minute. (The alternating current completes a cycle 60 times a second, and there are 60 seconds a minute.) Therefore, in a two-pole motor with the magnetic polarity making complete cyclic changes 3600 times per minute, the armature would normally try to make 3600 revolutions in a minute in an attempt to keep up with the magnetic field, as illustrated in Fig. 5-4.

In a four-pole motor, illustrated in Fig. 5-5, the armature, instead of turning a complete revolution, needs to turn only half a revolution before it meets the pole of opposite polarity. Consequently, it rotates only 1800 times a minute. A six-pole motor rotates 1200 times a minute, and an eight-pole motor rotates 900 revolutions per minute.

As we have seen, the armature attempts to keep up with the rate of magnetic polarity change, trying to rotate at the same speed as the magnetic field. The speed of the rotating field is known as the synchronous speed of the revolving magnetic field.

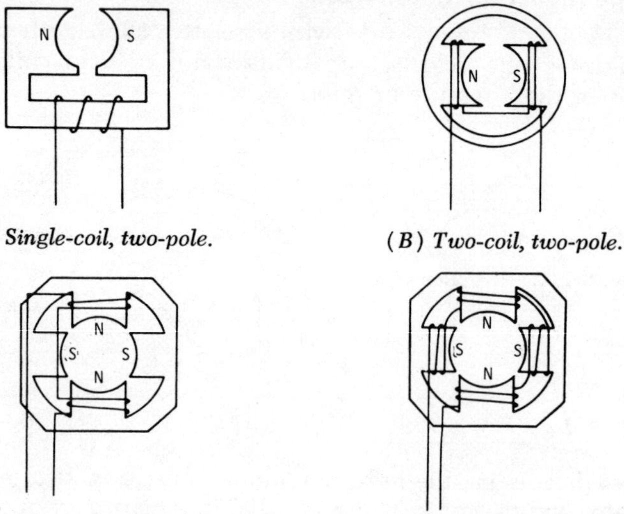

(A) *Single-coil, two-pole.* (B) *Two-coil, two-pole.*

(C) *Two-coil, four-pole.* (D) *Four-coil, four-pole.*

Fig. 5-3. Methods of stator construction.

There is a constant difference between the rotation of the magnetic field and the rotation of the armature. Consequently, the rotor speed can never reach the speed of the revolving field because it is necessary that the rotor bars continue to cut the flux of the stator field. The difference between the rotor speed and the synchronous speed of the revolving magnetic field is known as the "slip" of the motor.

The slip between the synchronous speed of the magnetic field and the rotation of the armature is just enough to create a current in the

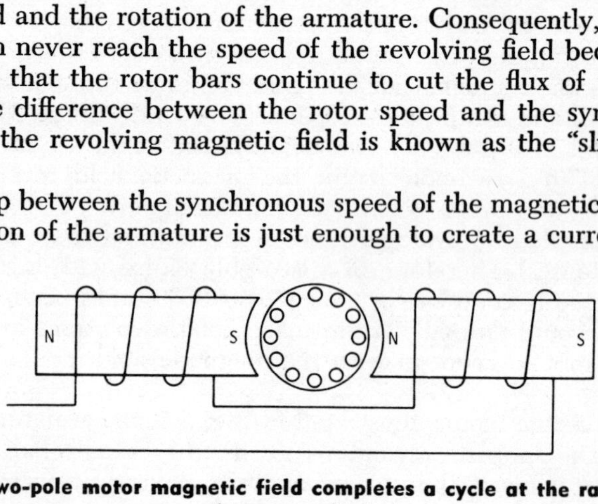

Fig. 5-4. Two-pole motor magnetic field completes a cycle at the rate of 3600 times per minute.

armature. That is, if the armature were to run at the same rate as the rotation of the magnetic field, no current would be created. However, the greater the difference between the speed of armature rotation and the synchronous speed of the revolving magnetic field, the greater the current and the greater the torque within the operating range of speeds.

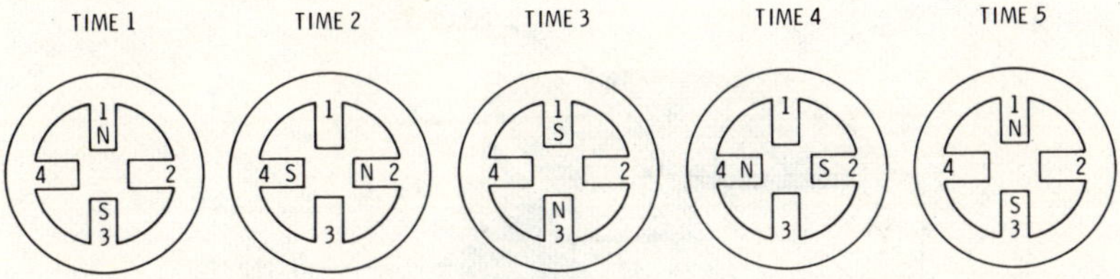

Fig. 5-5. A four-pole motor turns a half revolution before meeting a pole of opposite polarity.

The slip may be as high as 40% of the synchronous speed, depending on the load, the friction in the drive system, and the condition of the bearings. With increase in load, the speed is slower. Therefore, more flux lines are cut and greater torque is developed, depending on the speed-torque characteristics of the motor.

With an input line frequency of 60 hertz (cycles per second), the synchronous speed of a two-pole motor is 3600 rpm and that of a four-pole motor is 1800 rpm. Due to slip, the shaft speed of a two-pole motor will probably be within a range of 3000 to 3500 rpm and of a four-pole motor between 1400 and 1750 rpm.

Various means have been devised to start the induction motor and get it rotating. It can be seen that if the armature is standing still, the magnetic fields are simply reversing. Consequently, there is no incentive for the motor to turn. This necessitates placing additional poles in the motor structure which become energized at a slightly different time or out of phase with the applied voltage to the regular poles.

Single-phase induction motors are named and distinguished from one another according to the starting method of phase shifting. A phase shift in a winding can be accomplished by means of a series capacitor, by making one winding with more inductance than another winding, or, finally, by means of shading a pole.

Many tape-recorder motors use the shaded-pole principle. Starting torque is provided by pole shading, effecting a movement of magnetic flux across the pole faces of the field structure. This is done by notching each pole face and inserting a copper ring (called the shading coil) which encircles approximately one-third of the pole face as shown in Fig. 5-6.

The copper rings oppose a change in the magnetic field. Thus, when the main pole becomes north, the secondary pole created by the slot in the coils refuses to become north. Consequently, when the main pole is being reduced from north and is moving toward south, the slotted area becomes north, giving a rotational field which starts the motor.

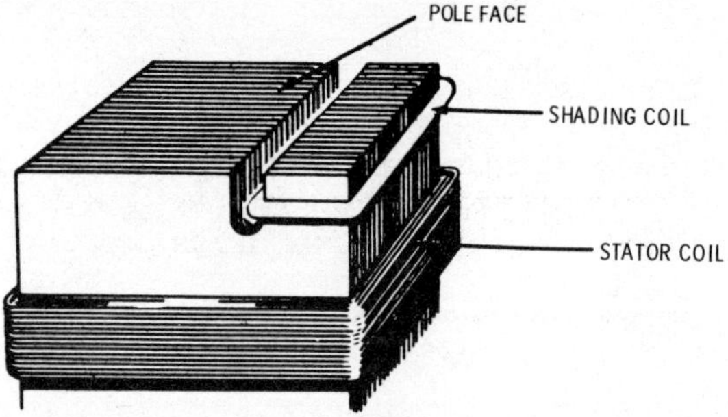

Fig. 5-6. Position of pole-shading coil in the pole face.

The four-pole motor is more satisfactory for use in tape recorders because it gives a lower external magnetic field. Two-pole motors have a tendency to induce a hum field in the magnetic head.

Other advantages of the four-pole motor include the lesser speed reduction necessary to drive the capstan, quieter operation is due to a lower vibration frequency, and the use of a larger rotor which can be dynamically balanced more easily, thereby further reducing motor vibration.

The four-pole, two-coil motor is seldom used in tape recorders due to its excessive external magnetic field which is almost as bad as the two-pole design. The four-pole, four-coil motor is more common.

SYNCHRONOUS MOTORS

The second type of motor used in tape recorders is the synchronous motor. In the synchronous motor the armature rotor travels at exactly the same rate of speed as the magnetic field. With the exception that the shaded-pole principle is not used (either the series capacitor or high inductance winding is common), the stator or pole structure is exactly the same as in an induction motor. The difference lies in the speed of the armature.

A synchronous motor, true to its name, runs at an exactly synchronous speed. If the magnetic field of the stator is rotating at 1800 rpm, the armature also will rotate at exactly 1800 rpm.

Synchronous armature speed is achieved by two methods. In the salient-pole synchronous motor illustrated in Fig. 5-7B, the armature is exactly the same as in an induction motor, using the same type of core structure that supplies torque. However, as shown in the comparison of the two armatures in Fig. 5-7A and B, the armature is milled with flat spots on it. As the motor comes up to speed, these flat spots will lock into the rotation of the magnetic field. Thus, an easier path for the magnetic field is supplied through one side of the armature than the other. The armature will tend to follow the magnetic field exactly rather than slip behind it, thereby running at synchronous speed.

The other type of synchronous motor, the hysteresis synchronous motor, is commonly used in present-day professional-type recorders. In a hysteresis synchronous motor, the armature is surrounded by a thin ring of magnetically hard material (Fig. 5-8) which becomes highly magnetized. As the motor comes up to speed, the rotating flux of the last cycle, before locking into synchronization, will magnetize one side of the armature north and one side of the armature south. These poles will lock in with the rotating magnetic field, and the motor will run at synchronous speeds.

The synchronous motor has the advantage that its speed will never vary. The speed of the hysteresis synchronous motor will be

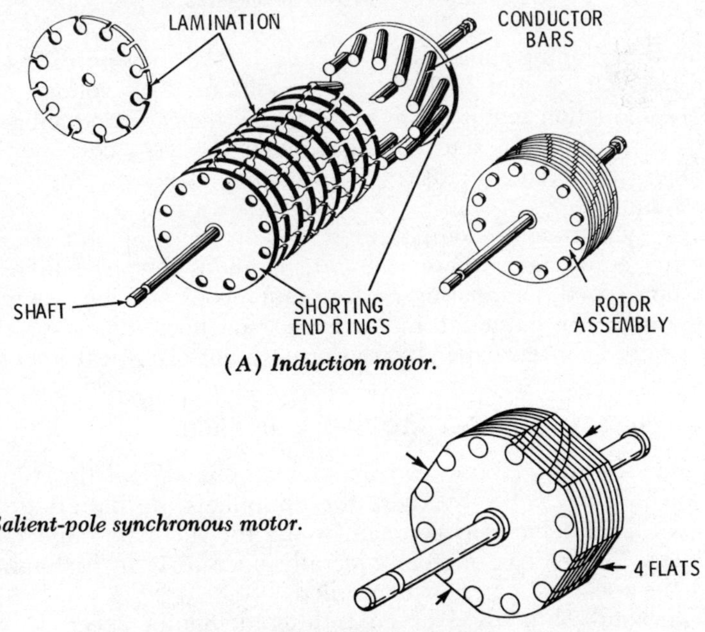

(*A*) *Induction motor.*

(*B*) *Salient-pole synchronous motor.*

Fig. 5-7. Construction of rotors.

as constant as the line voltage, and the recordings will last exactly the right length of time, providing all of the other drive-system parts are functioning without slippage.

INDUCTION VERSUS SYNCHRONOUS MOTORS

A synchronous motor tends, however, to have less desirable instantaneous speed characteristics. For example, the synchronous motor has trouble "hunting" or vibrating about a common speed, which introduces some flutter and wow into the recordings. On the other hand, the induction motor is not a constant-speed motor and

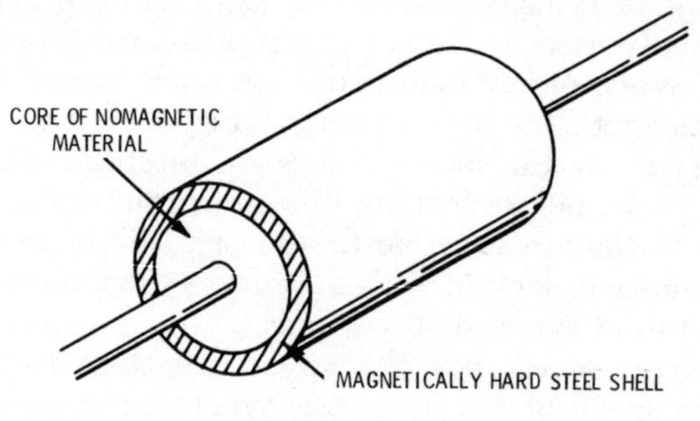

Fig. 5-8. Rotor of a hysteresis synchronous motor.

is always slipping somewhat. The rate of slip may be different under low-voltage and high-voltage conditions. The voltage applied to the induction motor changes its speed, hence introducing a timing error. However, the induction motor has very constant instantaneous speed characteristics, introducing little wow or flutter to the system.

The synchronous motor is most widely used on professional recording equipment where elaborate methods are used to insure a constant speed, eliminating both instantaneous and long-term speed variations. The induction motor is used on home-type tape recorders where long-term speed variations are not of critical importance.

DC MOTOR CONTROL

Since the introduction of transistor amplifiers and the miniaturization made possible by these tiny amplifiers, dc battery-operated motors have come into common usage for small portable tape recorders. These drive motors generally operate from flashlight cells and the speed is governor controlled.

The simplest method of controlling dc motor speed is by the variation of the input voltage. Speed may also be controlled by the output of the motor. A simple example of output control is the flyball governor used on steam engines. As the speed increases, the flyballs are forced farther from the center of rotation by centrifugal force. This movement, in turn, controls a valve which decreases the steam pressure to the cylinders. Using an output sensor to detect changes in speed and then controlling the input power is the original version of feedback control systems.

The simplest dc motor control employs a centrifugal switch which opens a pair of contacts as the speed of the motor increases above a certain preset level. When the centrifugally controlled contacts open, the current or voltage is reduced at the input of the motor, thereby slowing the speed to below the controlled rate. The contacts then close and full power is resumed at the input of the motor. An illustration of the basic method is shown in Fig. 5-9. By careful design, this make-and-break governor can be made to act within a very narrow range of speed variation (±0.1%) and at a high frequency rate.

There is one serious drawback to the use of the centrifugal switch to control current directly in magnetic recorders. Sparking at the contacts usually introduces noise into recording systems, and contact deterioration is excessive due to the frequency of the make and break.

This disadvantage is overcome by using the switch as a sensor to control the bias to a transistor which then handles the motor-control current. A circuit utilizing this principle is shown in Fig. 5-10.

The governor technique, as we said, is an electromechanical feedback system. If the circuit of Fig. 5-10 is subject to voltage fluctuations, an increase in voltage will momentarily increase current through R2 and increase the bias on the transistor XI. Increase in bias causes more current, the motor speeds up, and centrifugal switch C_s opens. Capacitor C1 charges, decreasing bias on Q1. Less current flows through the motor, it shows, and governor switch C_s opens.

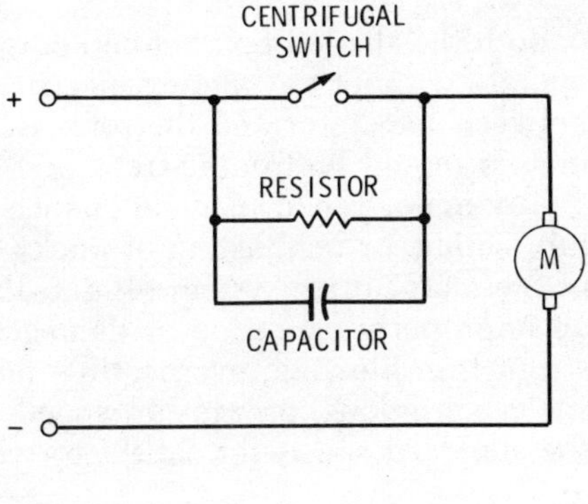

Fig. 5-9. Basic circuit of a centrifugal switch motor-speed control.

Thus, voltage changes will result in speed deviations before the stable speed is restored. In short, while this circuit can control speed to a predetermined standard for different steady voltages, it does not completely compensate for voltages that vary—for example, when a recorder is run from a car battery while the car is operating.

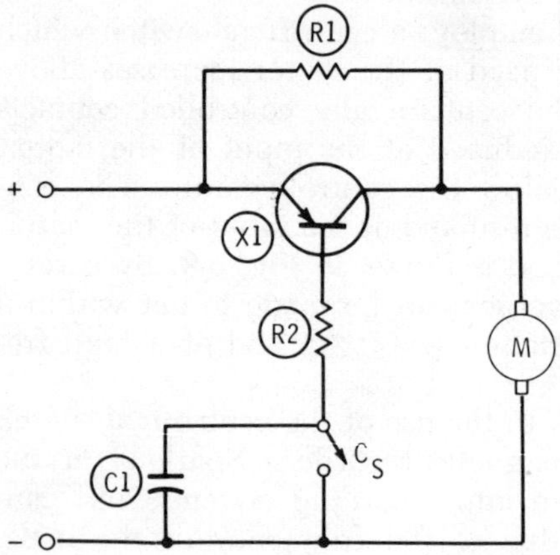

Fig. 5-10. Transistor motor-control circuit with centrifugal switch.

If the motor is well designed to run with low flutter, it will hold a constant speed when the voltage is held constant. A circuit that accomplishes this is shown in Fig. 5-11. The transistor is used in the emitter follower configuration, and the constant voltage at its base is provided by a zener diode, fed through an appropriate resistor. The capacitor is to bypass commutator ripple that otherwise might cause electrical interference in the recorder electronics (amplification).

MOTOR MAINTENANCE AND REPAIR

Both the shaded-pole induction motor and the synchronous motor operate without any sliding electrical contacts. The only connection between the stator and the rotor is magnetic. With normal use, the motors should be trouble-free.

Any motor repair or maintenance is generally limited to cleaning, lubricating, or readjusting of pucks, springs, bearings, or other parts of the motor drive system (Fig. 5-12). Faults that may be traced to an improperly operating or damaged motor or drive assembly consist of fuse blowing, overheating, mechanical noise, wow caused by speed variations, or capstan speed that is too far above or below the standard speed for satisfactory reproduction.

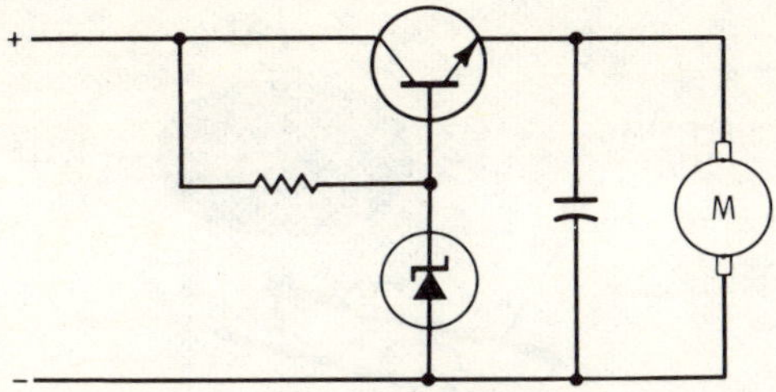

Fig. 5-11. A circuit that basis speed control on the supplying of a constant voltage.

Accidents or rough handling may break or short stator coils, bend the shaft, cause misalignment of parts, and break or loosen shading rings or shunts. Stator windings will burn out or short by the application of dc to 60-Hz motors. Overlubrication may result in slippage of the drive system due to oil on the idler. Lack of lubrication may cause bearing wear or freezing due to overheating when dry. Dirt, dampness, lubricants that gum, and accumulations of dust are enemies of moving parts.

If the capstan does not operate, fuses blow, or coils are overheating, check first for mechanical overloading. With a rim-drive motor this can be done by spinning the capstan and the rotor shaft with the idler disengaged. Then test the idler for sticking. All three should turn easily and freely. If they do, the next step is to check the electrical circuit of the motor. Mechanical noise and speed troubles require a more detailed physical inspection of the entire motor and drive mechanism.

All parts, surfaces, and dimensions that can affect the speed are of a critical nature. Intermittent slippage of some part of the drive system or parts that are not absolutely concentric with their axes of rotation will cause wow.

A motor will inherently transmit vibration to surrounding objects unless adequate provision is made to "float the motor" so that it is not solidly mounted, cushioning its movement. Rubber vibration mountings or grommets, surrounding the bolts or bushings, are used to fasten the motor plate to the underside of the base plate of the motorboard as shown in Fig. 5-13. If the vibration mountings are under compression, they cannot accomplish their purpose. There should be a small gap between the rubber and the hold-down washer. With some motors a spacer is included which is placed over the bolt and inside the grommet, preventing any erroneous tightening of the bolt. It is also essential that the rubber be resilient. Grommets that have hardened, collapsed, or become oil soaked should be replaced.

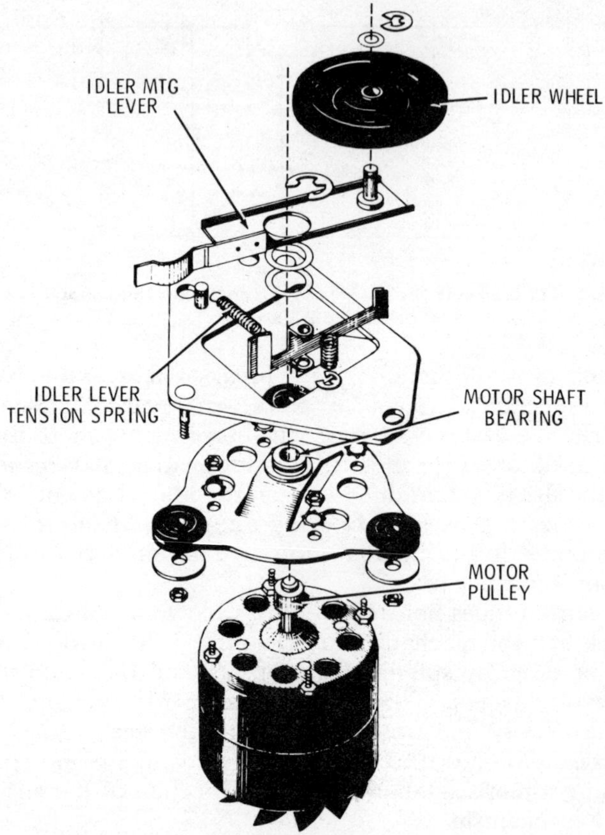

Fig. 5-12. Exploded view of a typical motor-drive mechanism.

Lubrication of the self-aligning–type rotor shaft bearings should seldom, if ever, be necessary. Motor manufacturers' advice concerning lubrication varies in detail. Nevertheless, all manufacturers caution against overlubrication and insist that any lubricant used

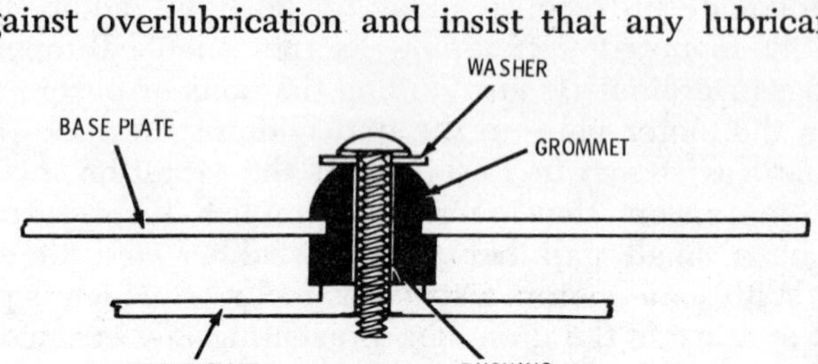

Fig. 5-13. A vibration mounting inserted between the motor plate and base plate.

must contain an oxidation inhibitor. It is also important that the viscosity remain nearly the same over a fairly wide temperature range. An oxidation inhibitor prevents the tendency of the oil to thicken, become gummy, or turn to a varnish under high-temperature conditions or long exposure to air.

Do not use any light household oils to lubricate motor parts, since these may not contain an inhibitor to prevent gumming. This risk can be avoided by buying regular motor oil that is known to contain an inhibitor. SAE No. 10 motor oil of high quality generally is satisfactory.

Shaft bearings that are gummy from being lubricated with a non-inhibited oil will need to be cleaned with a solvent before being lubricated. Apply only one or two drops of oil to the felt bearing washer, and after it has become saturated, wipe off thoroughly any excess oil that has reached other parts of the motor, giving special attention to the motor pulley and idler tire.

The air gap between the rotor and the stator is very narrow. To test for worn or misaligned bearings, check the shaft for play (Fig. 5-14). If the shaft can be moved at right angles to its axis, the rotor

Fig. 5-14. To test for misaligned bearings, check the motor shaft for play.

will, when revolving, be scraping the internal circumference of the stator laminations. Sometimes alignment of the sleeve-type bearings can be regained by tapping the bearing bracket. A typical motor using sleeve-type bearings is shown in Fig. 5-15. Tap the side of the gap that is narrowest with a light hammer, preferably a wooden or rawhide mallet. If the fault is due to the bearings being considerably worn, the motor or its bearings should be replaced. An exploded view of a typical end-bracket and self-aligning bearing assembly is illustrated in Fig. 5-16.

Ball-bearing motors are designed so that the bearings have a light press fit on the motor shaft and a sliding fit in the bearing housing. Adjusting the tightness of the case screws may be necessary with a ball-bearing motor before proper freeness is attained.

Fig. 5-15. A typical motor using sleeve-type bearings.

Worn or misaligned bearings, or a bent rotor shaft, will cause capstan speed variations by varying the friction of the motor pulley against the idler drive wheel. When the motor is running with the idler wheel removed, the appearance of the rotor shaft viewed from the end of the motor pulley should be circular and not a "wobbly" or elongated pattern.

The laminations, shading rings, magnetic shunts, bearing brackets, and screws which hold the motor to its mounting plate must be tightened securely to prevent buzzing. For this reason, the laminations are usually varnish impregnated and are firmly fastened by rivets while being subjected to a pressure of several tons.

Rivets that are loose should be tightened or driven out and replaced with new rivets or bolts. When tightening the bearing brackets, care should be taken that the alignment of the rotor and shaft are correct. Loose or broken shading rings may be soldered or welded and clinched. Broken shading rings will also cause a lowering of the speed under load. Buzzing shunts should be pressed out, bent slightly, and replaced. Due to the vital role of these bridges in

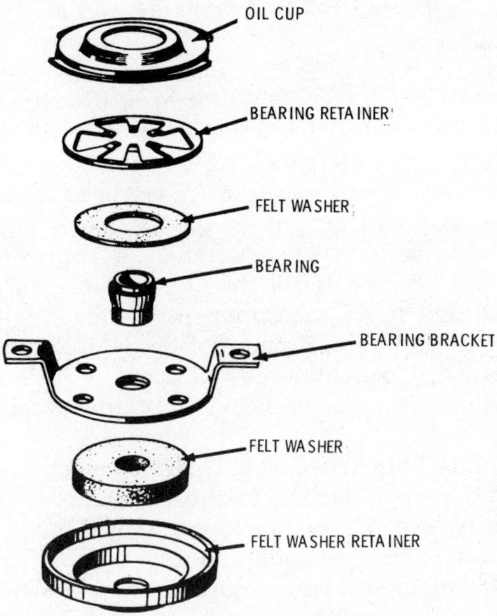

Fig. 5-16. Exploded view of typical end bracket and self-aligning bearing assembly.

motor performance, care should be taken to replace them in the same position they originally occupied. These bridges will sometimes be found with off-center holes or slots that affect motor operation.

If the capstan does not rotate, the motor is not stalled, and all parts are rotating freely, the electrical circuit should be investigated. Continuity tests should be made from the line input point through the switch to the motor coil. The resistance of the windings will probably be between 25 and 50 ohms. Opens external to the coil itself may be repaired. Connections between coils on motors of more than one coil must be kept in their original order. Incorrect connections will result in reversal of the fields which, with a two-pole, two-coil motor, would cause excessive current drain and no torque and, with a four-pole motor, would cause a great reduction of speed and torque.

Shorted turns in the coil will affect motor speed and can ordinarily be detected by the coil overheating and the unusually high current drain. It is not advisable or economically practical to try to rewind burned-out coils or those that have shorted windings. Since most of these motors do not employ layer windings, a special type of winding machine is required. Repairs necessitated by mechanical wear and strain can often be made reliably and at less cost by replacement of the motor.

When selecting the range for an ammeter to be used in a motor circuit, keep in mind that the power factor of the motor is apt to be low. Therefore, generally, the amount of current calculated by using the voltage and wattage rating of the motor should be approximately doubled. Also, motor ratings are usually given for "running" condition; the starting power will be much higher. For the small skeleton-type motor a current range of approximately 0.3 to 0.8 ampere should be expected if the motor is in good condition. It is recommended that initial measurements be made with an ac ammeter of 1-ampere full scale. Because of the relatively low efficiency of these motors (approximately 30%), there is little difference in the power demand between the heavy-duty motors and the smaller ones.

In most ac units hum will result if the connection from chassis ground to a solder lug attached to the laminations by one of the mounting bolts is broken. An ohmmeter should be used to check that the winding is not grounded to the laminations. Grounded windings would, in most cases, result in fuse blowing. They also may cause hum in the output.

Not much can be done through design to improve the efficiency of either the shaded-pole induction motor or the synchronous motor. As far as operating cost is concerned, it is relatively unimportant that there is a waste of power in the motor because the total amount required is so small. It is of more importance that the inherent inefficiency results in the watt-loss being converted into heat. Design limitations, choice of materials, and manufacturing cost considerations tend to further reduce efficiency to such an extent that the operating temperature of these motors is generally as high as safety permits.

Underwriters' Laboratories specifications for motors in recorder applications permit a coil temperature, as measured by thermocouple, of 194° Fahrenheit when the room ambient temperature is 77° Fahrenheit with the motor in normal operation in the unit, and with the maximum rated voltage and frequency applied. The temperature of the coil windings may rise to 221° Fahrenheit when resulting from a heat source external to the coil. Temperatures should not exceed 347° Fahrenheit under abnormal conditions such as locked rotors.

Shaded-pole motors are for alternating current use only, usually for 60-Hz frequency. Replacement motors for 50-Hz operation are available from most manufacturers. Consistent with the manufacturers' recommendation, conversion from 60-Hz to 50-Hz operation can sometimes by accomplished by changing the motor pulley to one of a larger diameter. This pulley must be supplied by the manufacturer of the recorder, and the conversion is possible only if the laminations and coil have been designed for use on both frequencies.

It is also possible to convert the recorder from 120-volt 60-Hz ac to dc operation by the addition of a vibrator or transistor inverter. Several of these small power supplies are on the market and are suitable for recorder use.

In the other direction, which is even more feasible, many of the newer portable transistorized cassette recorders that work primarily from batteries may easily be converted to a 120-volt ac supply with the addition of a small adapter that converts the ac supply to dc. This is accomplished by using a small transformer with rectifiers and a large-value, low-voltage electrolytic capacitor. Operation of the motor at the correct speed is usually no problem here. Whatever provision was made for achieving constant speed on battery operation also functions on the substitute dc supply.

CHAPTER 6

Handling the Range of Levels

The dynamic range of a system is limited by an upper and a lower level. As we commented earlier, the high noise level of primitive magnetic recorders hindered wide range recording. Improvements in materials and production methods have made possible the compression of more program time into an incredibly small space (from the viewpoint of earlier efforts) and the expansion of range.

The advantages to be gained by using better materials and techniques must always be shared between these two objectives—squeezing more programming into less space and achieving better quality, particularly dynamic range, in the recorded and reproduced program material.

An obvious step in this direction is to fully utilize whatever range you have. The program should be recorded at the maximum level the tape will accept without running into distortion. If it is recorded at a level lower than this, dynamic range is sacrificed needlessly, and the playback sound will seem noisier than necessary.

For this reason, most recorders have recording level meters. A variety of such indicators are now available. But to understand them, we need to look at the nature of program material, relative to this problem.

CAUSE OF DISTORTION

In magnetic recording, distortion is caused by the fact that as a given signal becomes progressively stronger and stronger, eventually the tape magnetism will no longer increase. The tape is then said to be *overloaded.* The overload point of a magnetic tape is also called *saturation* of the tape. One obviously wants to record considerably below tape saturation to avoid distortion. Fig. 6-1 shows the effect on a complex wave when recorded at too high a level.

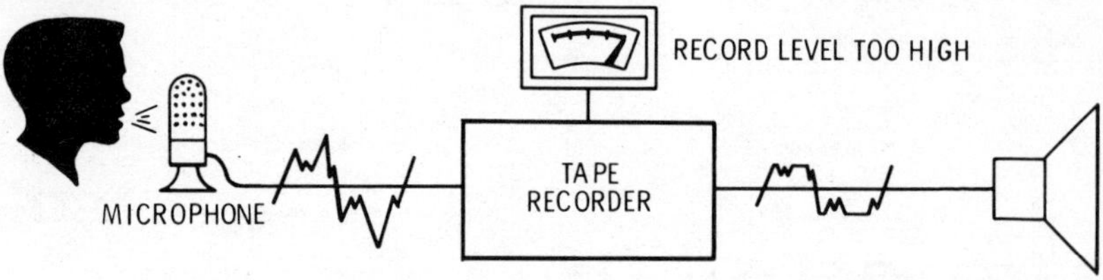

Fig. 6-1. Overrecording causes distortion on playback.

Tape records symmetrically; that is, when the tape leaves the erase head and before it passes across the magnetic record head, it is essentially neutral or in a demagnetized condition. As a signal is recorded on the tape, the tape is magnetized first in a north-pole direction and then in a south-pole direction. Since magnetization occurs symmetrically, there is no even-order harmonic distortion from the tape. However, when too strong a signal is applied to the tape and tape saturation is encountered, the peaks of the waveform are clipped. Serious third-harmonic distortion results, as shown in Fig. 6-2.

The normal method followed in adjusting the recording level of the tape is to record so that volume peaks never exceed 3% third-harmonic distortion. When we speak of peaks in recording music or voice, we are referring to the fact that the waveform is very irregular. At times the amplitude will be very weak. At other times it reaches very loud or electrically strong proportions as shown in Fig. 6-3A.

When recording very strong passages, care must be exercised to avoid overloading the tape as illustrated in Fig. 6-3B. On the other hand, do not record with too low a level so that weak passages are swallowed up in noise during playback as in Fig. 6-3C. Therefore, it can clearly be seen that the recording level determines the signal-to-noise ratio and the distortion. The prime rule to follow is to record at as high a level as possible without entering into the tape distortion region.

RECORDING VOLUME INDICATORS

To avoid the pitfall of entering the tape distortion region, some type of volume indicator is required to permit adjustment of the recording volume or signal amplitude as it is being fed to the tape. Over the years a number of different types of recording indicators have been tried. On present-day tape recorders there are three main types: the neon bulb, a peak-indicating device; the electron eye, again a peak-indicating device; and the VU meter, an average-indicating device. These three types of recording indicators are shown in Fig. 6-4.

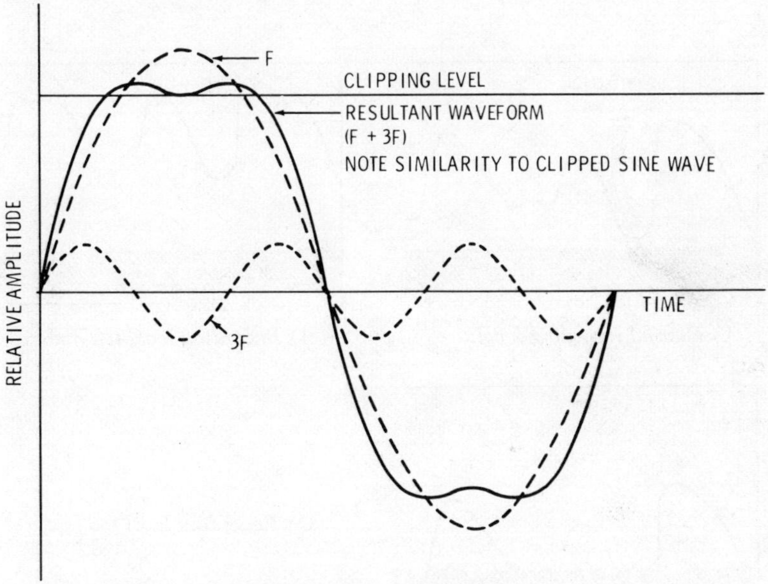

Fig. 6-2. When a sinusoidal wave is recorded at too high a level, serious third harmonic distortion occurs.

Becoming increasingly popular for the small portable recorders is a midget version of the VU meter. For many years, the "regular" size, shown in Fig. 6-4C, measured from 3 to 6 inches across. Many recorders now use sizes around 1 inch across; an even smaller size with no numbered scale as such is about 3/8 inch in diameter.

The Neon-Bulb Volume Indicator

The neon-bulb volume indicator is the most common type of volume indicator used in home-type recorders. A neon bulb is used which will not light until its applied voltage reaches a certain point; then, suddenly, it will ignite and glow. Thus, it can be used as a peak or volume indicator.

A peak is reached whenever the applied voltage exceeds the ignition or firing point of the neon bulb, causing it to light. Consequently, in the less expensive recorders using one neon bulb, the volume-level indicator will be adjusted so that when the signal reaches approximately 3% distortion (or generally a little less to provide a safety margin), it will light.

When recording, it is necessary to adjust the volume so that the bulb will only occasionally light. This means that during most recording the volume is adjusted well below the distortion point, with only the peaks entering the distortion region.

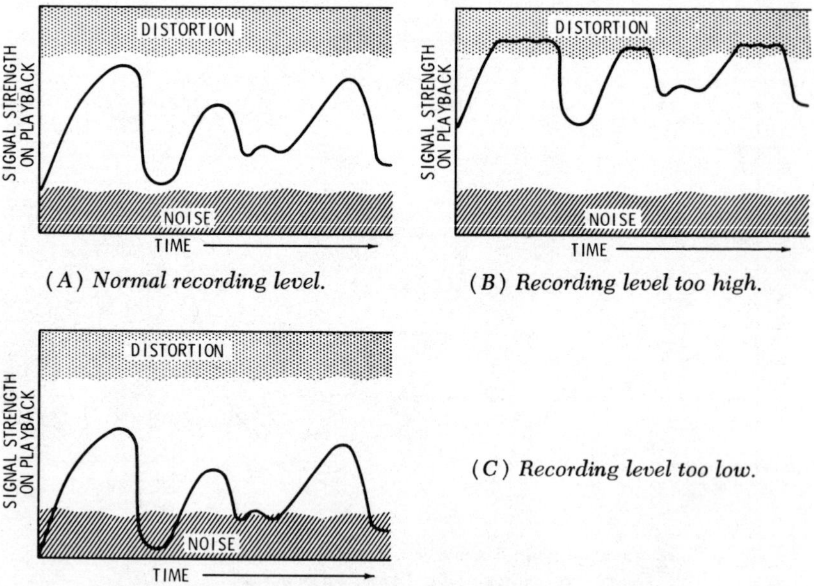

(A) *Normal recording level.*

(B) *Recording level too high.*

(C) *Recording level too low.*

Fig. 6-3. Normal recording level should be as high as possible without entering distortion region.

A further refinement of the neon-bulb indicator is accomplished by using two neon bulbs. For example, one neon bulb is set to fire when the recording signal reaches 3% harmonic distortion, whereas the other bulb is set to fire at a point somewhat under 1%. Consequently, the latter bulb serves as a lower-level volume indicator and the first bulb serves as a peak volume indicator. During recording one neon bulb is lit most of the time, indicating a normal recording level. When the second neon bulb fires, the peaks are entering the distortion region. The two-bulb system is a basic improvement and is incorporated on several of the newer home-type recorders. A partial schematic of a typical recorder using two neon bulbs is given in Fig. 6-5.

(*A*) *Neon-bulb indicator.*

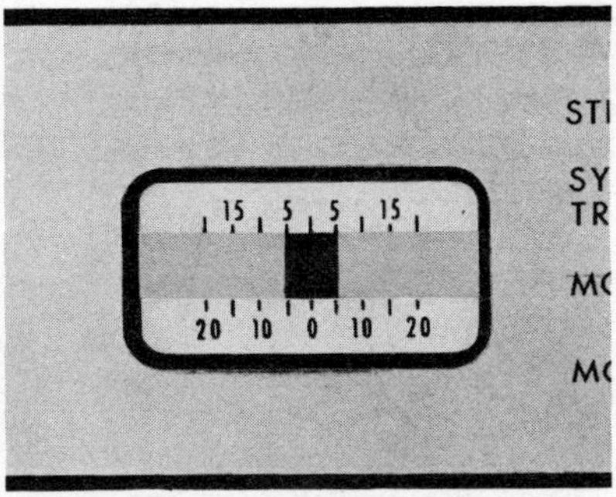

(*B*) *Electron-eye tube.*

(*C*) *VU meter.*

Fig. 6-4. Three types of recording level indicators.

The Electron-Eye Volume Indicator

The second volume-indicating device is the electron-eye tube. This tube has an electron beam that strikes a fluorescent surface, causing an image to appear. In the peak or fully modulated position, the electron-eye tube is as shown in Fig. 6-6. During recording, the two halves should be vibrating so that they never overlap in the manner shown in Fig. 6-7.

The electron eye has an advantage over the neon bulb in that a visual indication of the recording level is provided, even during weak signals. By contrast, in the one-bulb neon system, weak signals are not indicated, making it impossible to judge the amplitude of the applied signal.

Both the neon bulb and the electron-eye tube suffer from several disadvantages. The neon bulb will respond only to a given voltage, then will suddenly light, serving only as a peak indicating device. The electron-eye tube responds linearly to the applied voltage. That

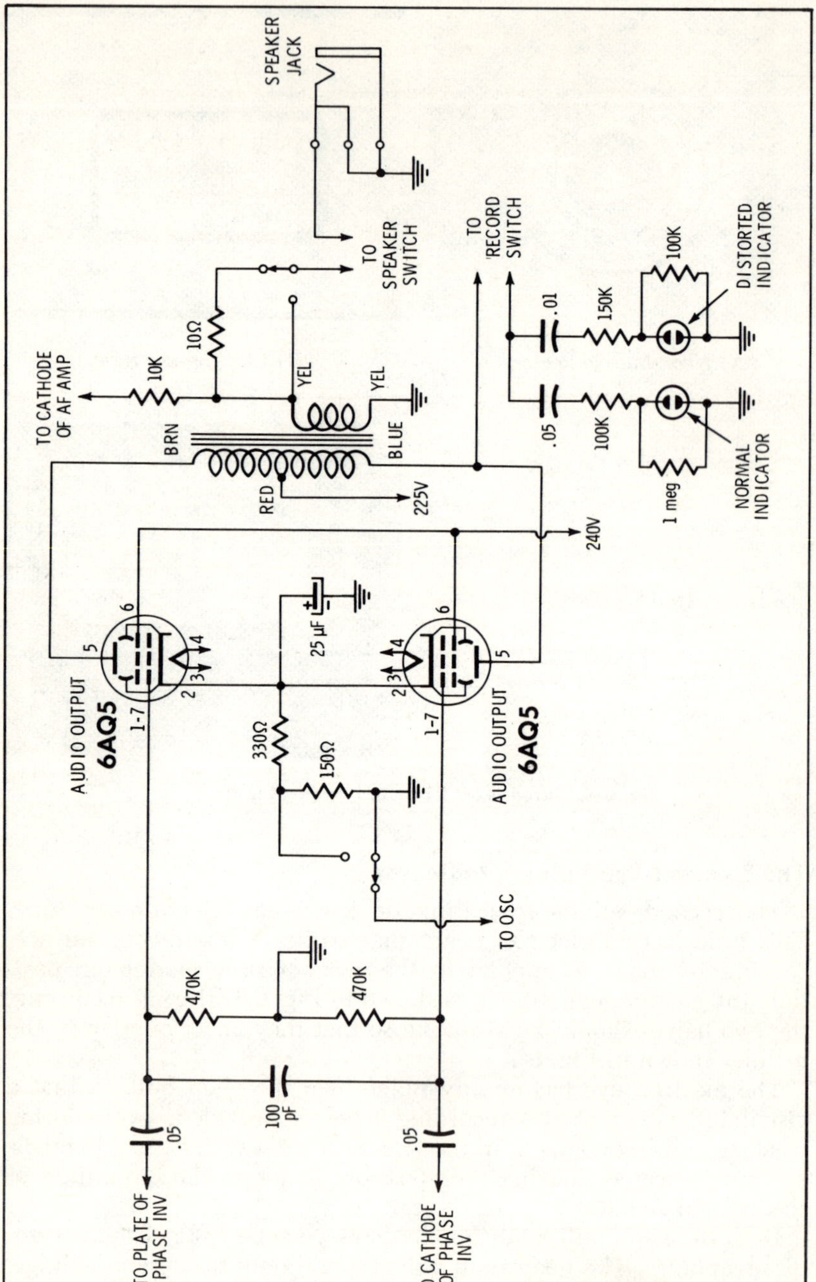

Fig. 6-5. Typical neon-bulb indicating circuit.

Fig. 6-6. Electron-eye tube in fully modulated condition.

is, it does not follow our hearing characteristics, which are logarithmic. In effect, this means that weak signals are completely lost or are barely indicated on the electron eye. Relatively strong signals are required to get a volume indication.

To avoid these disadvantages and to give an exact logarithmic indication of both weak and strong signals, a third device is used on practically all professional recording systems—the VU meter.

The VU Meter

The VU meter is an electrical meter responding in a logarithmic manner to an applied voltage or current. The VU meter is also an average-indicating device. A certain inertia is built into the needle

Fig. 6-7. Electron-eye indication of overrecording.

so that on a varying signal input, it will respond in a damped manner which gives an average indication of the signal level.

All VU meters are carefully built to have the same damping factor. Consequently, a recordist accustomed to reading one particular VU meter may look at a VU meter on any other recorder and judge the correct recording volume. Thus, the chance for error when switching from one recorder to another recorder is greatly reduced.

Since "0" level on a VU meter is an average and is considerably below the peak amplitude of the recording signal, it is necessary to locate the "0" level somewhat below the distorted region of the tape. Generally, "0" level on the VU meter, or 0 VU, is 8 dB below 3% harmonic distortion. This 8-dB margin gives ample protection for loud passages or loud peaks that are too rapid to show on the meter.

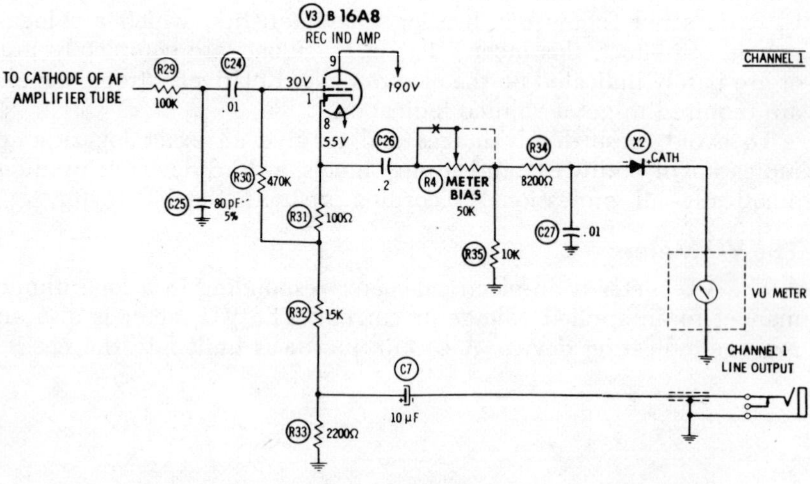

Fig. 6-8. VU meter with associated amplifier used in a typical more sophisticated recorder.

A circuit using a VU meter with an associated meter amplifier as used in the Concord Model 880 is shown in Fig. 6-8. This configuration appears twice in the complete stereo recorder-amplifier: one in the channel 1 amplifier and a duplicate in the channel 2 amplifier circuit.

The most common cause of trouble in a VU meter is that the needle becomes bent or the movement is actually burned out or damaged. Without exception, if the VU meter is damaged it should be sent to the factory for repair or replacement.

TESTS FOR PROPER ADJUSTMENT

A series of simple tests will quickly determine whether or not the volume-indicating device of a recorder is properly adjusted and is functioning correctly.

The neon-bulb indicator is generally connected across the output of the recording amplifier, just before the signal is fed into the recording head. Its reference level is established by applying a dc voltage to it. For example, if the bulb is intended to fire whenever the signal amplitude goes beyond 5 volts, and the ignition point, or the voltage required to light the tube, is 60 volts, 55 volts dc should be applied to the tube. Then the signal will be superimposed on the direct current so that whenever the signal amplitude reaches 5 volts the tube will fire. A voltage-versus-current curve for a typical neon bulb is given in Fig. 6-9.

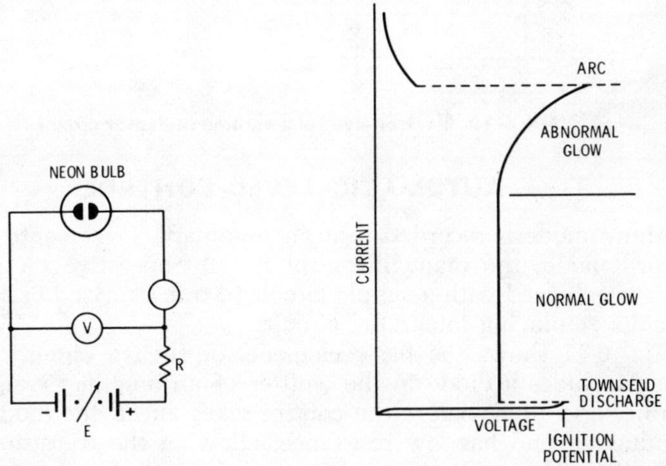

(A) *Method of measuring.* (B) *Volt-ampere characteristics.*

Fig. 6-9. Voltage-versus-current curve for a typical neon bulb.

All adjustments are usually fixed on most recorders. The resistances which determine the applied dc voltage are generally fixed values so that the neon bulb will light at a predetermined signal level. However, it can be checked with a distortion-measuring device. The neon bulb should fire at somewhere between 1% and 3% third-harmonic distortion.

Slightly more complicated than the neon-bulb indicator, the electron-eye tube gives the voltage instantaneously as the audio signal fluctuates. Since the audio signal fluctuates so rapidly, one would normally expect to see just a blur. However, in order to make readings possible, a capacitor network is used to remove rapid fluctuations and pass only gradual voltage changes, tending to give an average value of output somewhat similar to a VU-meter damping effect. A partial schematic showing a typical electron-eye circuit is given in Fig. 6-10.

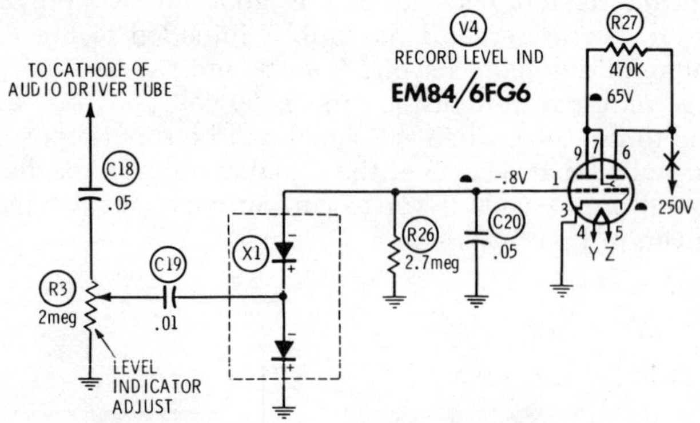

Fig. 6-10. Electron-eye tube volume indicator circuit.

AUTOMATIC LEVEL CONTROL

Many modern recorders use an automatic level control on the record mode; the manual control is only operative on playback. This is achieved with a simple circuit in transistorized or solid-state circuits employing integrated circuits.

Fig. 6-11 shows the basic elements of such a circuit. An early stage includes a diode in the emitter of an amplifier transistor. In normal operation, before the control takes effect, the diode is fully conducting and has low resistance, allowing the transistor to produce its full gain.

When the output signal is large enough to exceed a safe, undistorted level, a rectified signal is fed back to this diode in such a polarity as to reduce its conduction current. This raises its resistance value, and in consequence produces degeneration of the transistor's stage gain. With the reduced amplification, the output level is held within an undistorted range, provided the input is not more than the system can handle.

The time constant of the rectifier and its filtering action to the output signal are such that, after gain has thus been "turned down" by increasing the diode's resistance, the voltage slowly disappears if the output is not sustained, allowing the diode's resistance to drop again.

The effect of such a system is to maximize gain in the absence of program signal to record, and to minimize it in the presence of program signal. If, as often happens, such a recorder is used to interview someone in a location where there is some background noise (not necessarily loud), the effect can be disconcerting on playback. Any time there is a lull in the conversation, the gain will increase, making the background noise highly audible. When one of the parties to the interview speaks, the gain will turn down again and the background sounds will virtually disappear.

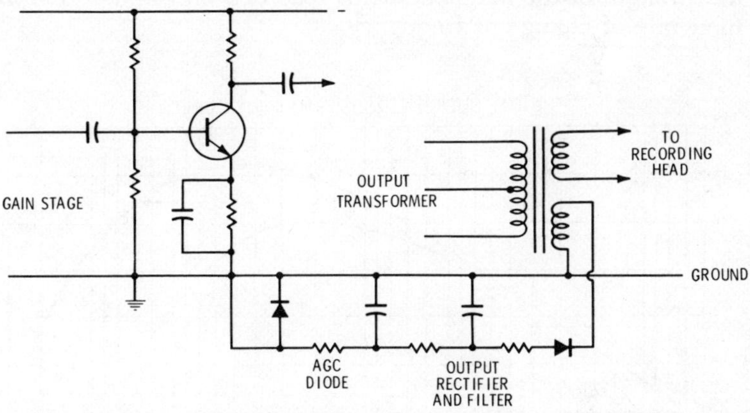

Fig. 6-11. Basic elements of automatic level control.

If such a recorder is used between two people with widely differing voice levels, one louder and one softer, the result can be disconcerting too. The automatic level control will compensate, but often not quickly enough. When the person with the softer voice responds after the person with the louder voice has been speaking, the former's words will be almost inaudible at first and will gradually "turn up." The remedy in this situation is either to use more than one mike and adjust the distances of the mikes so the pickup level is nearly the same or else use one mike and place the people at respective distances from the one mike. A good interviewer will watch for this effect and make changes accordingly.

If an experienced interviewer is conducting a discussion with several people, he will try to hold the microphone further from those who talk more loudly and closer to those who talk more softly. This is not always easy, as often people who talk softly tend to back away from a mike, while those who talk loudly try to get as close as possible to it.

EQUALIZATION CIRCUITS

Another consideration in regard to proper level is the level at different frequencies. When a constant amplitude signal is fed to a magnetic recording head and then played back, you might expect to obtain a flat or uniform amplitude over the frequency band. In reality, the resulting curve is far from uniform and flat, as shown in Fig. 6-12. Even though a constant-amplitude input was fed to the tape in Fig. 6-12, the output curve is insufficient at both the low and high frequencies. The reason for this is that in magnetic recording the voltage induced in the playback head is proportional not only to the strength of the magnetic fields recorded on the tape but also to their rate of change or frequency.

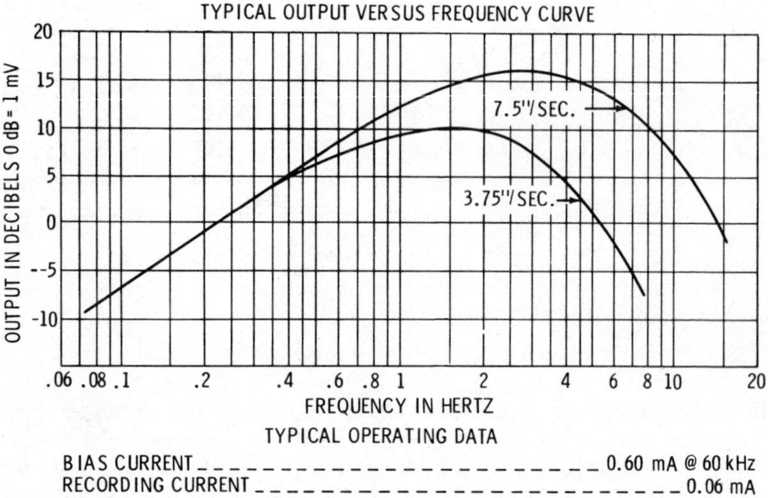

Fig. 6-12. Output versus frequency curve for a typical reproduce head recorded with a constant-current input.

Since a high-frequency signal has many more magnets and its field is changing faster than a low-frequency signal, it will produce a greater amplitude. In practice, a 200-Hz signal will give exactly twice the output voltage of a 100-Hz signal. That is, if the frequency is doubled, the output in magnetic recording is also doubled.

A simple computation shows that by doubling the voltage, a 6-dB increase in amplitude is obtained. Therefore, the magnetic-recording characteristics rise at the rate of 6 dB per octave. If the frequency is raised one octave or doubles, the output increases 6 dB.

This accounts for the falling off at the low frequencies. Around

1000 to 2000 Hz, the output reaches a peak and then tapers off at higher frequencies. This tapering off is due to losses in the head due to gap effects and the magnetic-head core structure.

NEED FOR EQUALIZATION

Because losses occur at both the low and high frequencies, some means must be provided to obtain a good frequency response on a magnetic recorder. This means is provided by equalization circuits which compensate for the losses at both the low and high ends.

The equalization may be accomplished by a variety of methods. However, in high quality home-recording equipment, the tape is usually recorded with a boost in the high frequencies, as shown in Fig. 6-13, and played back with an amplifier that boosts the low frequencies, as shown in Fig. 6-14. Thus, the low-frequency loss is corrected in playback and high-frequency loss is corrected in recording.

The reason for equalizing in this manner is to obtain the best possible signal-to-noise ratio without entering the distortion range. To obtain the best signal-to-noise ratio, it is desirable to put as strong a signal as possible on the tape, regardless of frequency, without overloading at any frequency. This means that if the tape

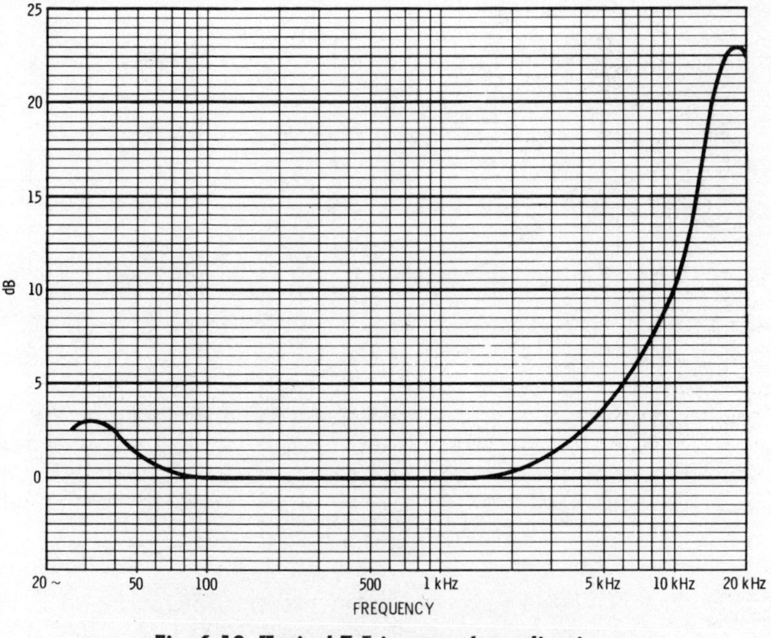

Fig. 6-13. Typical 7.5 ips record equalization.

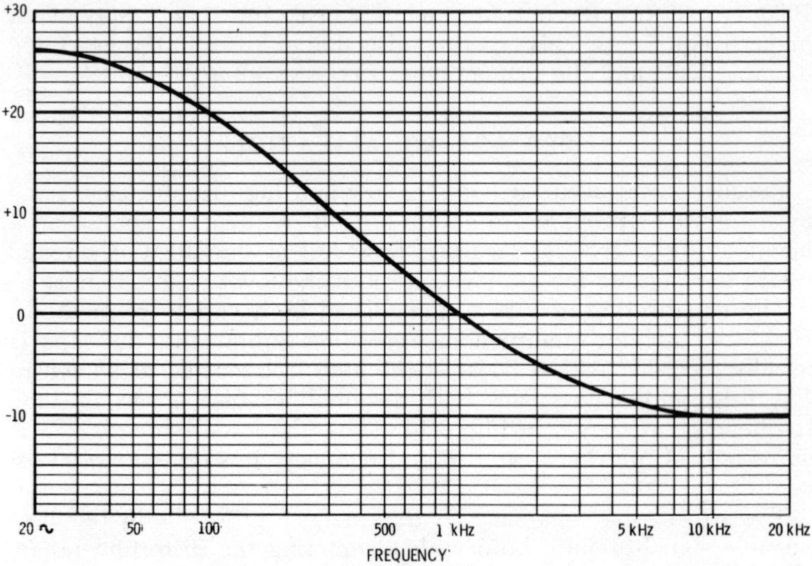

(A) NAB standard playback curve for 7.5 ips.

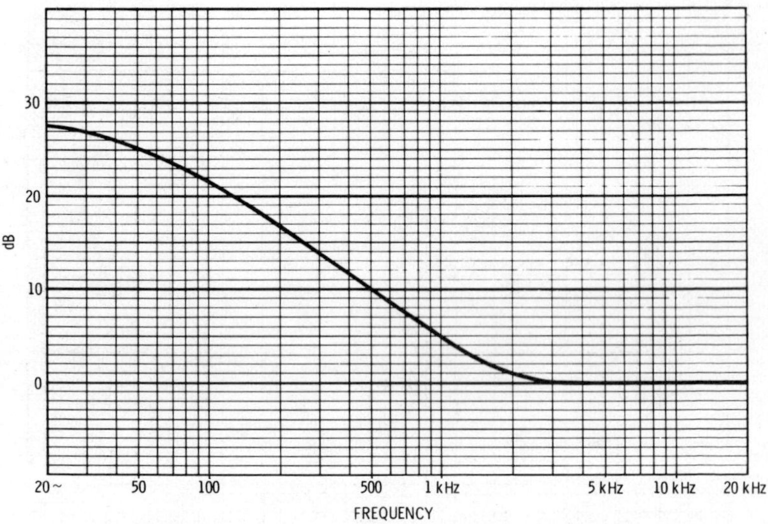

(B) MRIA suggested playback curve for 3.75 ips.

Fig. 6-14. Playback equalization compensates for record curves.

has a strong signal recorded on it, only a minimum of amplification will be required during playback.

The noise level in magnetic recording is generally due to tube hiss, microphonics, and hum created in the playback amplifier and does not result from the tape itself. In theory, tape has no surface noise, but has only modulation noise.

High-Frequency Equalization

If you record with a "flat" record amplifier (constant gain at all frequencies), the high frequencies are very weak when reproduced from the tape due to the losses at high frequencies, and excessive gain is needed to reproduce them at their normal level. However, boosting the high frequencies during the recording process will compensate for the losses, making it impossible to overload the tape. This is because sound waves generated by the human voice or a musical instrument have less strength at high frequencies, and the tape overloads at essentially the same amplitude regardless of frequencies. Therefore, high-frequency equalization can be accomplished during the record process without danger of the signal entering into the distortion range.

Low-Frequency Equalization

Low-frequency equalization is another matter. The power centered in vocal and musical sound waves is located in the low-frequency range. Consequently, low-frequency equalization must be accomplished during playback to avoid overloading or distorting the tape.

In less expensive recorders, however, low-frequency equalization is often accomplished in both the record and playback processes. While this does not theoretically give the best signal-to-noise ratio, in practice this method yields the best results on lower-priced machines. The reason for this is that in small, inexpensive home-type recorders, components such as the motor, power transformer, and amplifier produce large hum fields. Without expensive shielding, the magnetic reproduce head tends to pick up from these fields an excessive amount of hum.

If the full low-frequency equalization is accomplished in playback, the hum will be greatly amplified. However, if part of the low-frequency equalization is accomplished during the record process, less amplification will be needed at low frequencies during playback, thereby helping eliminate the hum.

Still other recorders, in an effort to simplify the record-playback switch functions, use only one equalization circuit for both record and playback. Therefore, both low frequencies and high frequencies are equalized in recording and playback.

The equalization used in the record process is referred to as *pre-equalization.* The equalization used in the playback process is known as *post-equalization.*

Differences in Equalization

Because of earlier differences in the equalization techniques employed by different recorder manufacturers, tapes recorded on one machine would not necessarily sound good on another machine. For example, consider a tape recorded on a professional recorder with no low-frequency pre-equalization but using the full high-frequency pre-equalization. The recording would sound unduly sharp and harsh with shrill highs and very little bass when played on a home-type recorder designed to have both low- and high-frequency equalization take place in the record and playback process. However, in modern machines, even the least expensive, the standard equalizations are used although the less expensive machines may not use precision tolerances.

EQUALIZATION FOR DUAL-SPEED RECORDERS

The graph in Fig. 6-15 illustrates that if the tape speed is doubled the region of peak output is raised one octave or 6 dB. When designing an equalization system for a dual-speed recorder, compensation must be made for the 6-dB-per-octave loss at the low frequencies and for magnetic-head gap and core losses at the high frequencies. To obtain maximum performance in dual-speed recorders it is imperative that an equalizer be designed for each speed.

To accomplish peak performance on dual-speed professional machines, two separate equalizer networks are provided. Only the pre-equalization is changed in several professional recorders. The post-equalization or low-frequency equalization is the same for both 7.5- and 15-ips speeds. By changing only the high-frequency equalization, flat response is obtained at both speeds. In other dual-speed professional recorders both the high- and low-frequency equalization is changed to compensate for the change in speed.

In many home-type, dual-speed recorders, the amplifier circuits embody only one equalization circuit. This circuit may be designed in three ways: First, the equalization circuit may provide for optimum results at one speed only. In the case of most home recorders which operate at both 3.75 and 7.5 ips, the equalization circuit is designed for the higher speed or 7.5 ips. This means that at 7.5 ips, optimum performance is obtained from the machine.

However, when the speed is switched to 3.75 ips, not only is response lost through the decrease in speed but far from optimum results are obtained since the equalization circuit is not designed for the 3.75-ips speed. Consequently, when a tape is played at 3.75 ips on an equalization circuit designed to compensate for losses at 7.5 ips, the results cannot help but be disappointing.

The second approach sometimes used in dual-speed home-type recorder equalization circuits is to provide equalization for a nonexistent intermediate speed. As an example, the equalization circuit may give optimum results at a speed of 5 ips. Therefore, when a

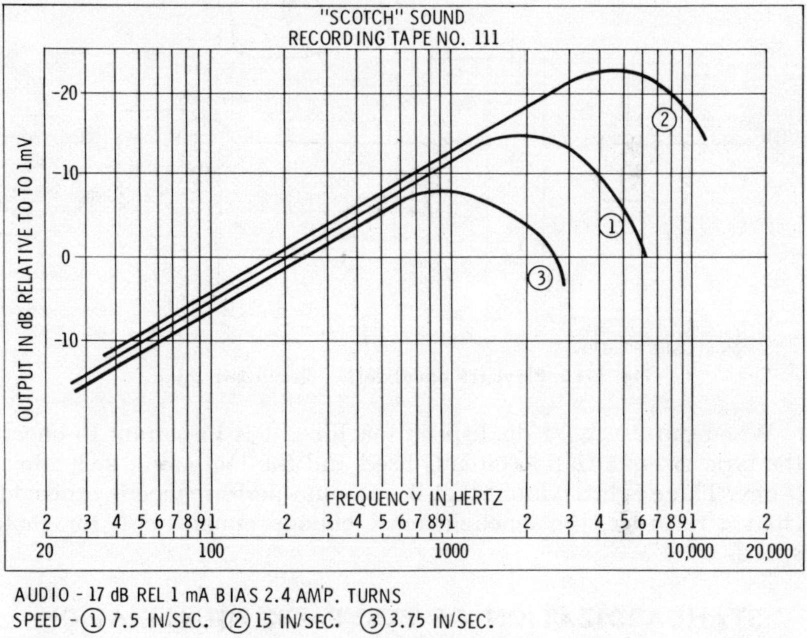

Courtesy 3M Company

Fig. 6-15. Effect of speed on frequency response.

machine is operating at either 3.75 or 7.5 ips, the results are naturally not as good as could be expected. True, the 7.5-ips speed will give better results than 3.75 ips. However, neither speed will be entirely satisfactory. A frank compromise, this type of equalization circuit is designed to improve the 3.75-ips response while sacrificing some of the advantages obtained with a 7.5-ips speed response.

A third method of equalization of dual-speed home-type recorders is finding increasing favor. This type of equalization recognizes the limitations in a short-cut, compromise approach and, instead, employs two different equalization circuits (Fig. 6-16).

Those recorders that use two separate equalization circuits generally switch between the circuits with the same knob that controls the speed. For example, when switching to 7.5 ips, the equalization is automatically adjusted for that speed.

Of course, separate equalizer circuits must be incorporated in each channel of a stereo amplifier. Thus, a unit which is capable of recording and playing back stereo and has separate equalizers for each speed would actually contain four equalizer circuits.

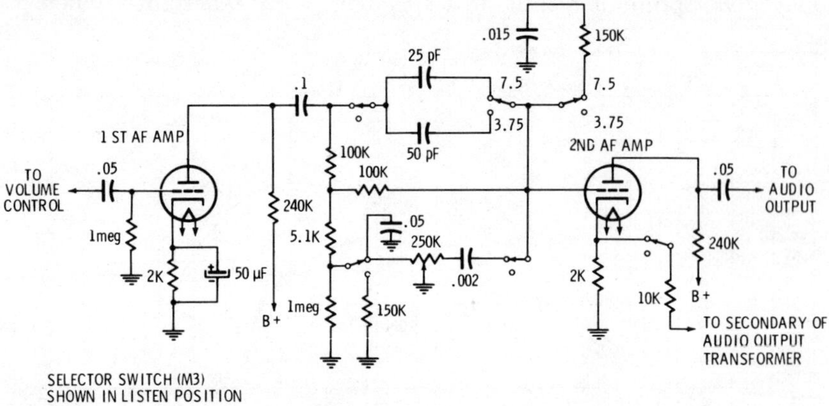

Fig. 6-16. Playback equalization circuit switching.

When purchasing a dual-speed machine, it is important to check the type of equalization circuits used and the method of switching, if any. There is little doubt that if optimum performance is expected from a recorder, the machine must change equalization for each speed.

STANDARDIZATION OF PLAYBACK EQUALIZATION

The National Association of Broadcasters (NAB) has adopted a standard professional playback-equalization characteristic curve. Tapes are recorded in such a manner that when played back with the standard playback characteristic, the output will be essentially flat. The NAB curve for 7.5 ips is shown in Fig. 6-17. This curve is also generally accepted for 3.75 ips.

The Magnetic Recording Industry Association (MRIA) has formed an aggressive standardization committee to work out equalization circuits among all American recorder manufacturers. In addition, those foreign tape-recorder manufacturers interested in exporting their machines into the American market have also been invited to participate. If tape is to compete on an even basis with present-day discs for the broad home-music market, standard playback curves must be used on home-type recorders so that recorded music on tape will sound equally good on all machines. It is ob-

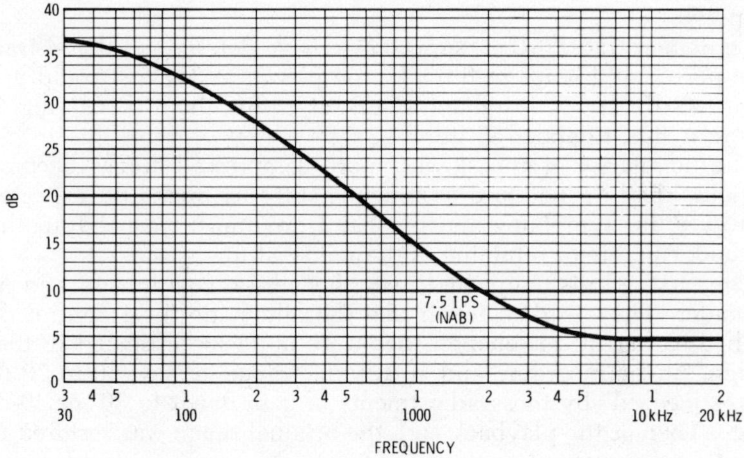

Fig. 6-17. NAB equalization curve for 7.5 ips.

viously an economic impossibility to expect dealers to stock a line of recorded tapes specially equalized for each of the major home-type recorders. This would be necessary for optimum results if a standard equalization were not used.

NOISE REDUCTION

Obtaining the proper level and fitting the frequency content into the available level range were only partial methods for accomplishing superb reproduction of original sound. As the pressure to record more program in less space led to slower tape speeds and narrower tracks, the difficulties of achieving a clear reproduction multiplied.

The main problem was that the top level, at which distortion occurs, did not have sufficient range above the lowest level, at which tape noise occurs. When a 2-track tape was played at 30 ips, the range between noise level and distortion (called the dynamic range of the tape) was from 50 to 60 dB, representing a voltage range of from approximately 300 to 1 to 1000 to 1.

Improving the tape quality and squeezing more tracks onto the tape made possible the use of speeds of 3¾ ips or even 1⅞ ips. However, the dynamic range was cut from 50 to 60 dB to a 20-, 30-, or 40-dB level. For many purposes these reduced ranges, or depths, are quite adequate. For dictation purposes, 20 dB is sufficient. For dialog, comedy routines, and small group popular music, 30 dB is probably adequate much of the time. But for symphony, big band and other more sophisticated kinds of program, there are times when even 60 dB is not enough dynamic range to "handle" it properly.

To demonstrate further, suppose we have determined that 4-track tape played at 1⅞ ips or 8-track tape played at 3¾ ips would give about 40 dB, at 7½ ips about 45 dB, at 15 ips about 50 dB, and at 30 ips, it rises to about 55 dB.

The popularity of the slower, more compressed tracks becomes obvious—they do not lose as much quality as might be expected. However, for symphony and big band programs, we need another method, capable of retaining high quality at low speeds.

The first efforts employed two devices—a compressor and an expander. They worked much like the automatic level control described earlier in this chapter, but were more sophisticated in their design. At the recording end, a dynamic range of, say, 60 to 70 dB was "squeezed" by these adjustments in gain down to 30- or 40-dB range. Then at the playback end, the original range was restored by inverting the process.

How did this work? Peaks that did not rise more than approximately 10 dB above the full range "floor" level received full amplification. For successively higher peaks, the gain was reduced by half the amount by which the peaks exceeded that level. Peaks going 30 dB above floor level were reduced to 10 dB, 50 dB were reduced to 20 dB and 70 dB were reduced to 30 dB. Thus the original 70-dB range was effectively compressed into 40 dB. Fig. 6-18 shows the system.

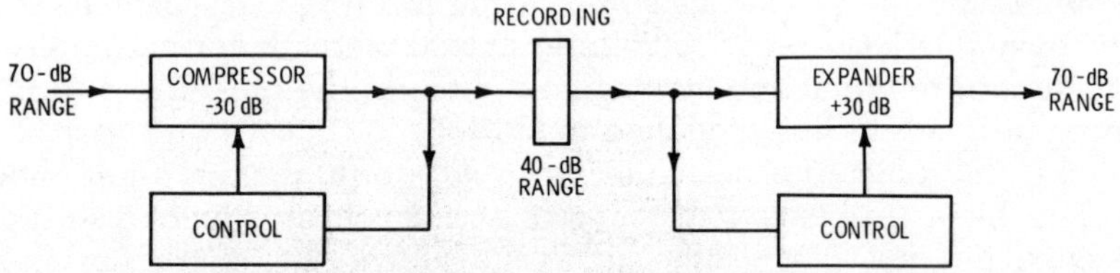

Fig. 6-18. Block diagram of noise-reduction system using an expander/compressor combination.

With good demonstrations these systems were convincing. But attempts to use them for everyday recordings and reproduction ran into difficulties. If levels were not correct everywhere, it was easy to "miss" so that either the top levels ran into bad distortion or the bottom levels were down in the noise. Some forms of programs proved susceptible to distortion, just as a normal result of being processed in this way. Periodic loud spots, such as cymbal clashes, could result in a sort of "breathing sound" as the background noise went up and down.

Second efforts produced the Dolby system, a two-step system that operates before and after recording. When the strength of certain components of lower-level signals being recorded falls below a predetermined set level, the system amplifies them. During playback, a mirror image of the same Dolby process that amplified these signal components before recording returns them to the original level. At the same time, the noise added during the recording system is reduced.

Use of the Dolby system in home recorders provides 3 dB of noise reduction at 600 Hz, 8 dB at 2 kHz, and 10 dB at 4 kHz.

Where the compressor/expander system pulls down the top level for strong signals—no matter what frequencies made them strong—the Dolby system raises weak signals to push them above the noise level. It does this on a frequency-selective basis, raising only the level of those frequencies that are weak.

CHAPTER 7

The Bias Oscillator

A jack of two trades, the bias/erase oscillator serves the all-important function of making linear and distortion-free low-noise recordings as well as erasing the tape. Unlike the proverbial jack of all trades, however, who was master at none, the bias oscillator performs both functions admirably.

TWO GENERAL BIAS METHODS

There are two general bias methods employed in magnetic recording. One is direct-current bias; the other is alternating-current bias.

The early wire and steel-band magnetic recorders of Valdemar Poulsen's day used direct-current bias. One of the chief limitations of the Poulsen magnetic recorder was the use of direct-current bias, which prevented the recorders from wide public acceptance. Despite Poulsen's brilliant design, the use of dc bias inevitably resulted in distorted recordings with considerable hiss and noise.

However, with the introduction of alternating-current (ac) bias in the middle twenties, magnetic recordings of high quality became possible for the first time. This discovery of W. L. Carlson and G. W. Carpenter of the U.S. Naval Research Laboratory opened up an entirely new vista for magnetic recording.

The purpose of alternating bias is to produce a linear recording characteristic. This means that if the signal is doubled, one will obtain twice the output from the tape during playback. If the recording characteristics were nonlinear and the signals were doubled, one might obtain either more or less than twice the output.

In the magnetic-recording process, the recording characteristics must be linear. In magnetizing any iron or steel substance, a hysteresis curve is encountered. A typical hysteresis curve is shown in Fig. 7-1. The horizontal axis indicates the strength of the recording signal; the vertical axis gives the amount of magnetic field left in the tape.

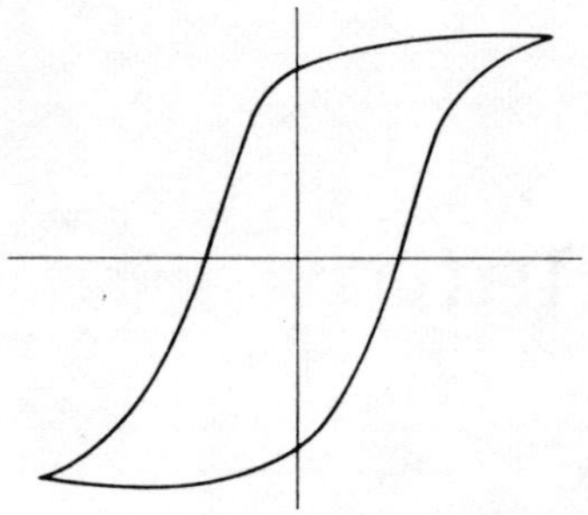

Fig. 7-1. Hysteresis curve of typical magnetic-recording material.

When an increasing magnetizing force is applied to the tape and then removed, the tape will have a certain remanent flux. In magnetic recording, the remanent flux is the actual magnetic field that creates the signal during playback. Ideally, if the tape were recorded linearly for a given magnetic force recording signal, an equal remanent flux would be obtained (Fig. 7-2). If the recording signal were doubled, theoretically, the remanent flux would be

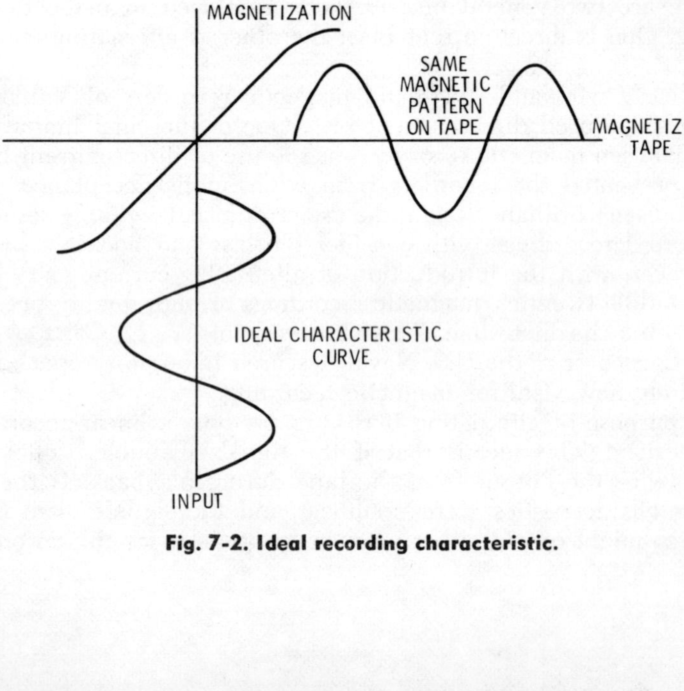

Fig. 7-2. Ideal recording characteristic.

doubled. However, from the curve shown in Fig. 7-3 it can be seen that the magnetic field on the tape is no longer proportional to the recording signal.

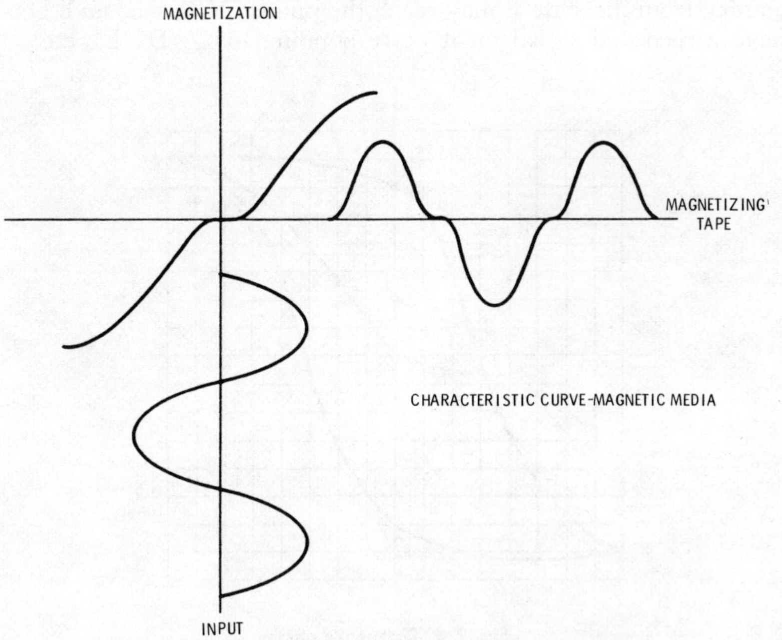

Fig. 7-3. Actual recording characteristic with no bias.

Theory of DC Bias

The first serious bias attempt was the dc bias method discovered and incorporated by Poulsen in his *Telegraphone* recorder. It is interesting to note how dc bias operated. In dc bias, the tape was completely magnetized by a permanent magnet, generally serving also as the tape eraser. Thus, when the tape met the recording head, it was in a completely magnetized condition. All the magnets were oriented in one direction. Then, as a recording signal was applied, the recording signal served to demagnetize the tape. The amount of demagnetization was determined by the strength of the signal. At the same time a certain amount of direct current was applied to the head. The direct current was used to place the operating signal within a more or less linear region of the hysteresis curve.

When recording is to be accomplished by means of dc bias, the procedure involved is the following: The recording medium is first of all magnetically saturated in a certain direction with a dc erase head, as indicated by OAB in Fig. 7-4. Then in the record head, the function of the dc bias is to bring the medium to a point such as D. In the absence of a signal, the tape leaving the record head will be in the magnetized state D′. In the presence of a signal, the tape coming from the state B may reach the points DCE, etc., and, hence have a recorded signal on it corresponding to C′, D′, E′, etc.

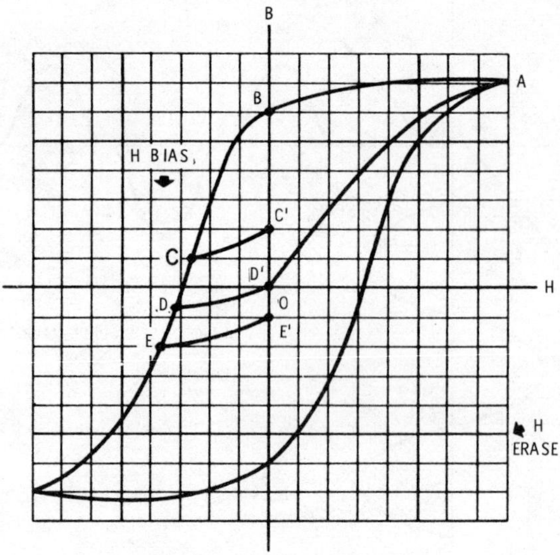

Fig. 7-4. Dc record curve.

When employing dc bias, very little of the potential magnetization of the tape was used, resulting in noisy recordings with a high hiss level. Unless elaborate precautions were taken to adjust the dc or permanent magnetic fields accurately, a large amount of distortion would also result.

The defects of the dc bias method led to the development of the alternating-current bias. Ac bias is now used in all magnetic recording applications for high quality audio work.

Frequency of the AC Bias Supply

It is obvious that an alternating current is needed to supply an ac bias system. The bias current is generally many times the strength of the recording current. Consequently, far more bias than audio signal is needed in the record head. The amount of bias may be as much as 10 times the strength of the audio signal, although generally it is only several times as great.

Fig. 7-5 illustrates the action of ac biasing. A field which is a mixture of the ac bias and an audio signal is shown applied to the residual flux-density curve. In Fig. 7-5 the projection of the positive peaks on the residual-flux curve at each instant of the audio signal produces curve A. The projection of the negative peaks forms curve B. The addition of curves A and B represents the effective-flux signal that will be reproduced. Although some distortion is present in the resultant curve, it is a great improvement over reproduction without bias.

In order not to be heard, the bias must be at a higher frequency than the audio signal. The frequency of the bias, while not critical, must be well above a certain value. This point is generally five times the highest audio frequency to be reproduced. This is necessary in order to avoid beats or whistles in the audio frequencies being heard during the recording process. The harmonics of the audio signal that beat with the bias frequency cause a whistle or heterodyne.

When a whistle develops, two alternating currents of different frequencies are mixed together in a slightly nonlinear recording system. For example, if an audio frequency of 10,000 Hz were recorded and a bias frequency of 35,000 Hz were used, the third harmonic of the audio frequency would be 30,000 Hz. Since the bias frequency was 35,000 Hz, a 5000-Hz intermittent whistle would be heard in the recording.

Generally the second- and third-harmonic components of an audio signal are the strongest. The fifth-harmonic component is rela-

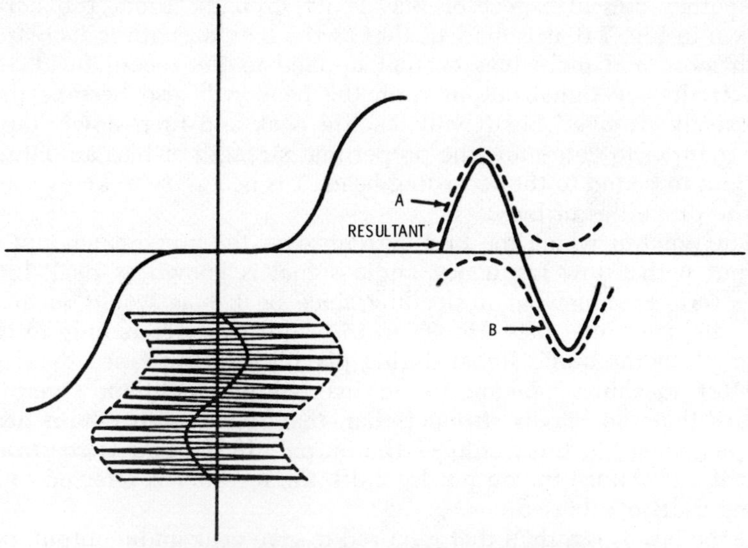

Fig. 7-5. Magnetic characteristics with bias.

tively weak, giving little trouble. Therefore, if the highest audio frequency to be recorded is 15,000 Hz, the bias frequency should be at least 60,000 Hz. Or, again, if the highest audio frequency to be recorded is, say, 7500 Hz, the bias frequency should be at least 30,000 Hz. The trend of recent recorder design is toward higher bias frequencies in order to avoid any chance of heterodynes.

In present-day professional recorder design, bias frequencies generally are in the region of 80 kHz. Home recorders use bias frequencies in the region of 40 to 80 kHz. A few of the older home recorders still operate at bias frequencies as low as 30 kHz.

Erasure

There is no limitation on how high a bias frequency can be used, except for those limitations imposed by the bias and erase currents.

Since the bias supply serves as a supersonic oscillator, it forms a handy source of power for the erase head during erasure of the tape. The erase head requires considerable power to completely erase the tape. With the trend toward higher and higher bias frequencies, it is becoming increasingly difficult to apply sizable amounts of power to an erase head. Therefore, it is desirable from a design standpoint to use a relatively low bias frequency in order to have sufficient power to operate the erase. However, soundwise it is best to have as high an audio frequency as practical.

Bias Strength

Another crucial aspect of bias is its strength. From the curve shown in Fig. 7-6, it is evident that as the bias strength is increased with more and more bias current applied to the record head, the low frequency signal output from the head will also become progressively stronger, but it will reach a peak and then slowly taper off. In order to determine the proper bias strength or bias amplitude current to be fed to the recording head, it is necessary to know some of the properties of bias.

The point at which the bias current gives the greatest amount of output with a low frequency audio signal is known as peak bias. This term is somewhat misleading since peak bias would seem to indicate a stronger bias. However, the word peak refers only to the strength of the audio signal during playback of the tape.

Most machines operate in an overbiased condition. Overbias means that the bias is stronger than that required to obtain peak output which, in turn, reduces the output. If the bias is so strong that it will reduce the output by 2 dB, the recorder is referred to as being 2-dB overbiased.

If the bias is less than that required to give peak audio output, not only will the output suffer, but the signal-to-noise ratio will also be

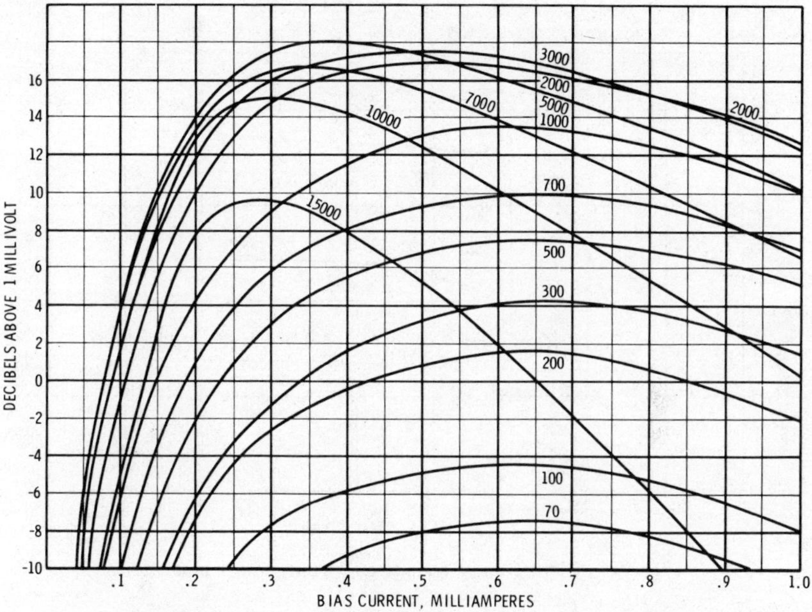

Fig. 7-6. Self-erasure effect of bias in high-frequency recording.

adversely affected, and severe distortion, together with poor uniformity, will set in (Fig. 7-7). Even the nodules and imperfections on the tape surface itself will be accentuated with low bias. However, the main disadvantage of low bias is the high distortion and low-signal output encountered.

In the case of overbias, the distortion is generally somewhat reduced beyond the point of peak bias. In fact, as the bias is increased the distortion becomes progressively less. However, the signal output begins to drop, though slowly, causing a reduction in the signal-to-noise ratio.

The main disadvantage of overbiasing is the marked loss of high frequencies (Fig. 7-8). As the bias strength is increased, the bias actually serves as an erasing field for the tape, erasing the weaker high-frequency signals on the tape. Since high-frequency signals are more easily erased than low frequencies, the bias is actually erasing the high-frequency audio signals on the tape. By overbiasing by one or two dB, the high frequencies will begin to be attenuated, although they are not severely affected with this amount of overbias. On an oscilloscope the surface uniformity will look better, and the distortion characteristics are generally improved.

Therefore, in summary, if the bias is too low distortion will result. If the bias is too high a loss of high frequencies will occur. The

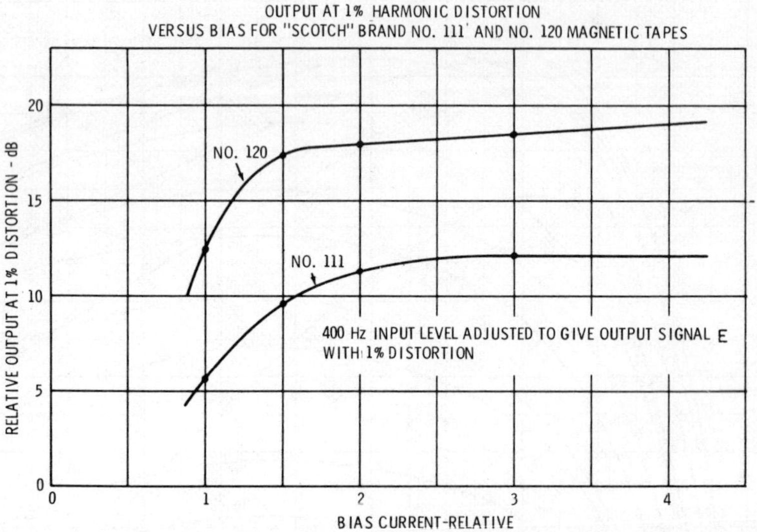

Fig. 7-7. Distortion in output as a function of bias.

decision, then, is this: Should a recorder be operated at peak bias or slightly over peak bias? Operation at peak bias usually makes possible the best high-frequency response since it is in a very good region as far as distortion is concerned. If distortion is very low,

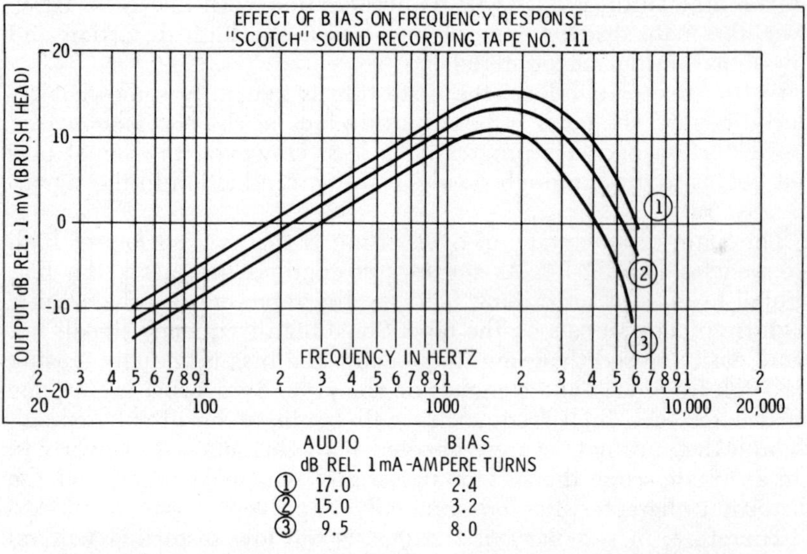

Fig. 7-8. Effect of bias on output for a constant low-frequency signal.

peak bias generally will also give the best signal-to-noise ratio. The only disadvantage of peak bias is that the surface irregularities of magnetic tape have a tendency to be somewhat more pronounced. However, thanks to recent improvements in magnetic-tape construction techniques, nodules and other surface irregularities are relatively unimportant as far as audio applications are concerned.

ADJUSTABLE AND FIXED BIAS

The philosophy of most machine manufacturers, if they do not incorporate an adjustable bias in their recorder, is to build machines fixed at somewhat overbias. This technique, due to variation in production components, will allow a few machines to be biased at peak bias while a few machines will also be substantially overbiased. However, most machines will fall in a region of approximately 2-dB overbias. In general, the performance of most machines can be improved if an adjustable bias is incorporated in their design. Most professional recorders have an adjustable bias such as that shown in Fig. 7-9.

BIAS-OSCILLATOR DESIGN

The bias oscillator itself is generally a simple LC circuit with either a tuned-plate or a tuned-grid class-C oscillator employing a triode tube. In some of the professional recorders, dual triodes are used in push-pull, as shown in Fig. 7-10, in an effort to reduce second-harmonic distortion. Many models now use transistor bias oscillators, similar to the one illustrated in the schematic of Fig. 7-11.

A common cause of distortion arises from an improper bias waveform. In the bias-signal wave, if one side of the wave has greater amplitude than the other, the tape will become magnetized and, as discussed earlier, second-harmonic distortion will be introduced. Therefore, it is extremely important that the bias supply have a good sinusoidal waveform. In an effort to do this, the professional manufacturer almost invariably will use push-pull oscillator tubes to eliminate any second-harmonic component in the bias waveform that could introduce second-harmonic distortion in the tape recording itself.

While very little bias current is needed for the recording process, the bias oscillator is also called on to supply considerable current for the erase head. The current required for erasure is often as much as ten to thirty times as high as the current required for the recording process. Therefore, an oscillator must have substantial power to deliver the current for erasure. Most erase heads need almost 4 watts of power to do a complete job of erasure. This means that the oscillator must be capable of supplying up to 4 watts of power at a high frequency. In general, this is the frequency limitation imposed on the bias oscillator.

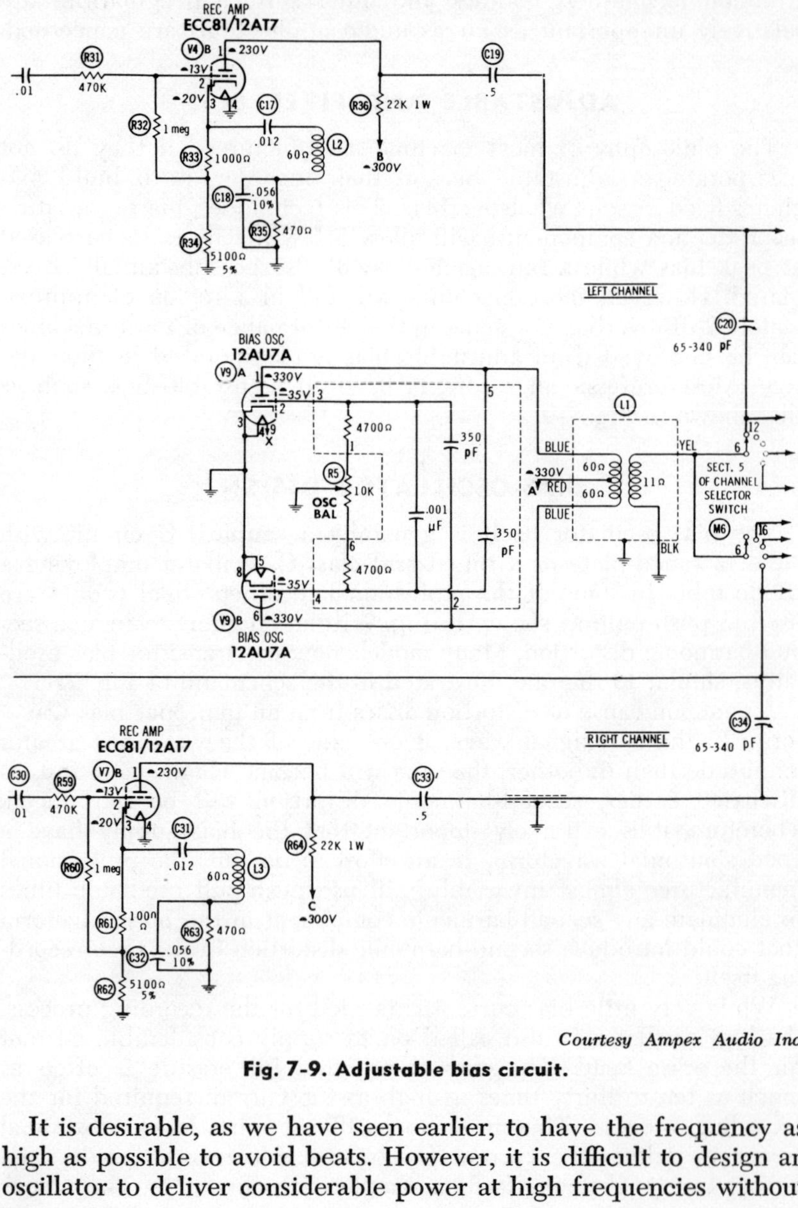

Courtesy Ampex Audio Inc.

Fig. 7-9. Adjustable bias circuit.

It is desirable, as we have seen earlier, to have the frequency as high as possible to avoid beats. However, it is difficult to design an oscillator to deliver considerable power at high frequencies without

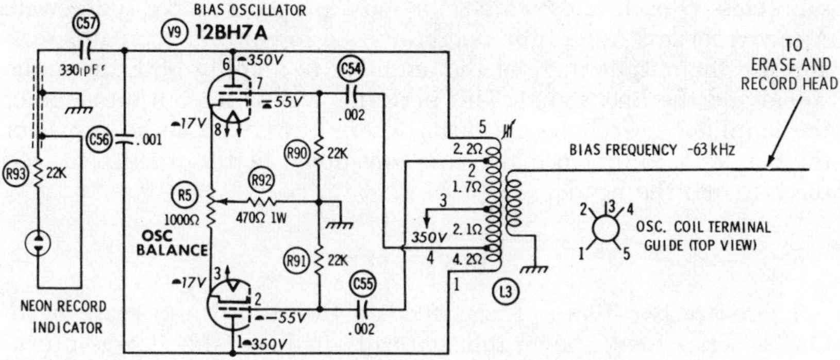

Courtesy *Midwestern Instruments, Inc.*

Fig. 7-10. Push-pull bias-oscillator circuit.

involving added expense. Therefore, the design of the bias oscillator represents a compromise between sufficient output for tape erasure and high enough frequency to avoid heterodynes in the recording signal. The lower-priced home-type recorders generally use single-

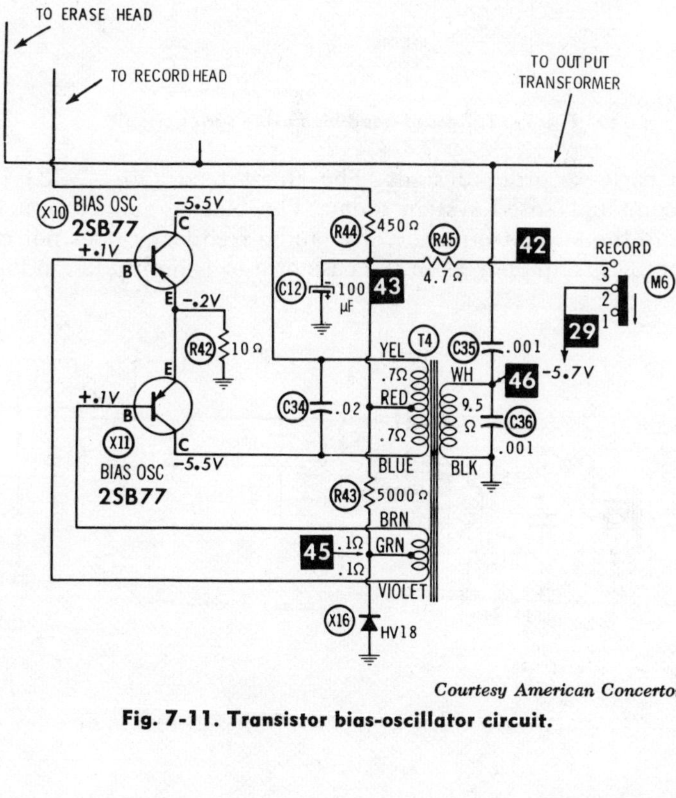

Courtesy *American Concertone, Inc.*

Fig. 7-11. Transistor bias-oscillator circuit.

tube class-C oscillators which, for most purposes, serve quite well. A few recorders use a tube operating as a low-power oscillator tube and use the output stage of the amplifier to amplify both the audio signal and the bias signal. This performs well if the output stage of the amplifier introduces no distortion in either the audio signal or the bias waveform. Normally, it is considered better practice to feed directly into the head.

BIAS FEED

There are two types of bias feed to the record and erase head. One is series feed, shown schematically in Fig. 7-12. It was preva-

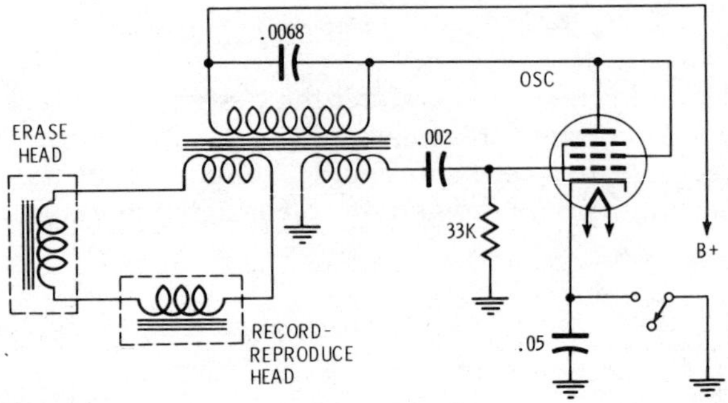

Fig. 7-12. Record-head bias using series circuit.

ent in early recorder designs. The shunt feed (Fig. 7-13) is the most commonly used system today. The bias frequency is actually added to the audio frequency. The audio frequency does not modulate the bias frequency as in the case of a-m (amplitude-modulated)

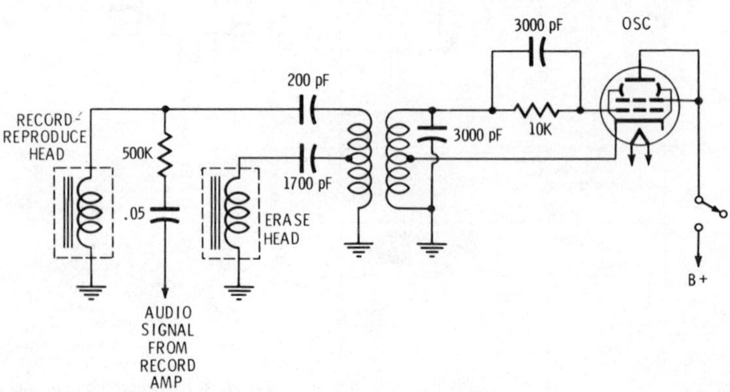

Fig. 7-13. A parallel-feed bias circuit.

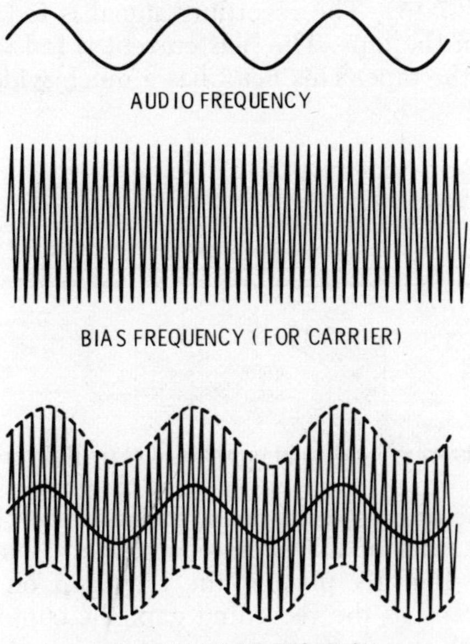

Fig. 7-14. Audio and bias are mixed before going to the record head.

radio waves. Instead, a waveform such as those shown in Fig. 7-14 is produced.

The series feed has the disadvantage that the bias is not easily controllable. If the bias on the record head is adjusted, either the erase current will be lowered or, if the bias is shunted through the record head, a load will be placed on the audio signal. Therefore, the series-feed method does not lend itself toward an adjustable bias control. For this reason, the series-feed has been generally abandoned.

In shunt feed, it is possible to regulate both the erase current and the bias current. The most common type of bias feed in use today, shunt feed, requires a simpler head design and permits the use of high-impedance heads.

THE CROSS-FIELD TECHNIQUE

The whole question of achieving optimum bias was for a long while one of compromise. A high value of bias reduces distortion, but also decreases high-frequency response. Insufficient bias produces distortion, poor signal-to-noise ratio, and reduced output.

A recent trend uses interacting heads to achieve biasing of the input signal. In one system, the heads are located on opposite sides of the tape (Fig. 7-15). The recording signal is fed to the head on the coated side of the tape. The bias current is fed to a head on the opposite side of the tape. This head has a much wider gap than the record head.

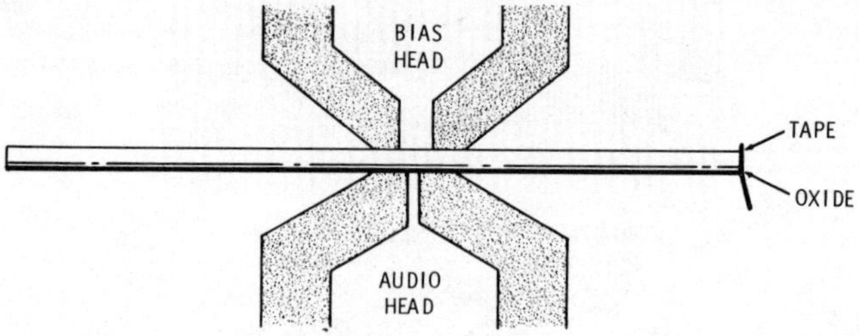

Fig. 7-15. An arrangement used to produce cross-field magnetization.

In another type of cross-field head, the function of the additional head, which is a part of the same composite assembly, is slightly different (Fig. 7-16). By placing the bias gap on the same side (coated) of the tape as the recording gap, the contour of the magnetic field through which the tape moves is changed. This contour change in the magnetic field results in a deeper penetration of the recorded magnetization into the tape, particularly the high frequencies.

TROUBLESHOOTING

If a recorder has severe distortion in the audio signal, one of the first places to look is in the bias oscillator. Is the bias oscillator functioning? Is it functioning well?

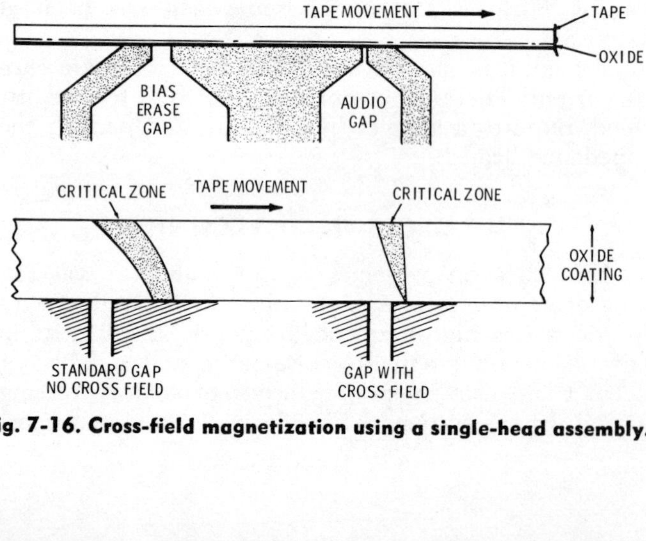

Fig. 7-16. Cross-field magnetization using a single-head assembly.

If the oscillator becomes at all weak, the erase current will suffer, and the bias will, of course, be reduced while the signal will become distorted.

If the tube or transistor is not faulty and the signal is still found to be weak, look for faulty components. The bias waveform should be checked with an oscilloscope to obtain a rough approximation of its shape—it can be easily seen if the output is a good sinusoidal wave. If the waveform is not sinusoidal, something is definitely wrong in the oscillator and should be corrected to obtain good, low-distortion recording results.

CHAPTER 8

The Record and Playback Amplifier

The signal from the microphone is extremely weak, at the most only a few millivolts. Therefore, during recording an amplifier is needed to raise the feeble signal from the microphone to a high level before feeding into the magnetic head. Adding still further complication during playback, the signal obtained from the magnetic reproduce head is also very weak, at the most approximately 8 millivolts. Again, an amplifier with very high gain is required to raise the signal strength sufficiently to drive a speaker.

Because of the weak signal from both the microphone and magnetic reproduce head, it is easy to see that the record and playback amplifiers play a crucial role in the tape-recording process. Both amplifiers operate from the same basic principles and have the same limitations. The only differences between the record and playback amplifier lie in the amount of gain required and the type of equalization used, a discussion which is covered in another chapter.

While the discussions in this chapter might seem to apply only to monaural units, it should be remembered that the only difference between stereo and monaural units is that a stereo unit will have two identical amplifier channels.

AMPLIFIER CHARACTERISTICS

As the name implies, an amplifier is constructed to amplify or build a weak signal into a strong one. That is, the amplifier takes a weak electrical signal, generally an electrical voltage produced by either a microphone or the signal from the magnetic reproduce head in the playback process, and raises this voltage many thousands of times. The signal is subsequently converted into current which is, in turn, used to drive a speaker.

The amplifying process or the amount of amplification is technically known as the gain of the amplifier. There are two types of gain: voltage gain and power gain. Supplying voltage gain is the principal function of both the record and playback amplifiers, except for the output stage in the playback amplifier where voltage is converted to power.

To obtain gain, a vacuum tube or transistor is employed. The function of a vacuum tube or transistor is similar to that of a valve; a relatively weak voltage is used to control a large current or voltage. If the gain from one stage is insufficient, the tubes or transistors can be cascaded. In cascading, the output from the first stage is fed into a second, the output from the second is fed into a third, etc. Then if each stage is amplified, say, 10 times, two stages would amplify 100 times, and three stages 1000 times.

In tape recorder applications, the gain of the first stage may be several hundred. The following stage usually has a gain of between 10 and 60. Therefore, in the first two stages of the cascade chain, there may be a gain of well over two or three thousand.

The gain of the amplifier is determined by three factors: first, the type of tube or transistor used; second, the values of the circuit components; and third, the mode of operation of the tube or transistor.

Type of Amplifying Device Required

In the early stages of a tape-recorder record-playback amplifier, a very high amplification is necessary to raise the weak signal to a satisfactory level while introducing as small an amount of noise as possible. In order to accomplish this objective, it is desirable to amplify the weak signal as rapidly as possible and then to amplify a relatively strong signal in the subsequent stages.

The pentode-type tube has an amplification of from 100 to 300 times. In this respect it is very desirable since it has high gain. However, it is noisier than the triode or transistor and introduces somewhat more distortion, although distortion is not a big problem in the early amplifier stages. A triode, while it has less gain, has far less noise and substantially less distortion. However, to duplicate the gain of the pentode, two triodes are necessary. The total results are often very similar, regardless of whether amplification is obtained with two triodes or with one pentode. The number of circuit components is nearly the same.

Later tube developments led to the adoption of low-noise tubes. Several excellent pentodes became available, and the trend in recorder manufacture went toward tubes like the EF86/6267, which gives higher amplification with less noise than former types.

The amplification at midfrequencies can be easily determined by the following formula:

$$A = \mu \frac{R_1}{r_p + R_1}$$

where,

μ is the amplification factor of the tube,
r_p is the plate resistance of the tube,
R_1 is the load resistance and the grid resistance of the following stage in parallel. That is:

$$R_1 = \frac{R_L \times R_g}{R_L + R_g}$$

where,

R_L is the load resistance,
R_g is the grid resistance.

The vacuum tube, whether it is a pentode or a triode, can be represented as a generator whose output is the grid signal, or the input voltage to the grid of the tube, times the amplification factor of the tube. In series with this generator is the plate resistance inherent in the tube (Fig. 8-1). By knowing these values, the circuit can be designed to give the desired gain while still possessing satisfactory frequency response.

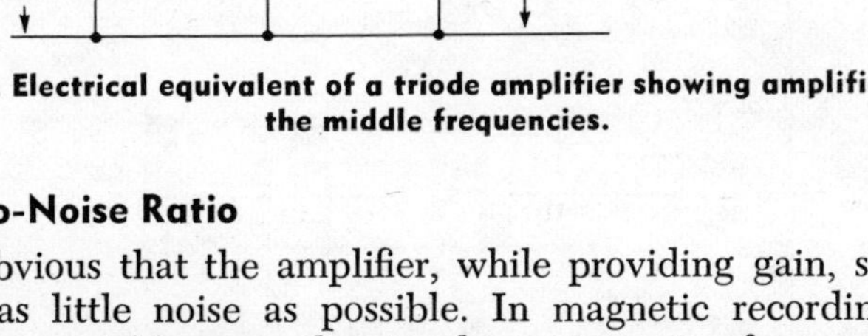

Fig. 8-1. Electrical equivalent of a triode amplifier showing amplification at the middle frequencies.

Signal-to-Noise Ratio

It is obvious that the amplifier, while providing gain, should introduce as little noise as possible. In magnetic recording this is especially crucial because the signal-to-noise ratio of a tape recorder is primarily limited by the playback amplifier.

Signal-to-noise ratio, simply stated, is the maximum signal obtainable from the amplifier before distortion sets in, compared to the noise inherent in the system. In tape recording, the medium of magnetic tape introduces no surface noise. That is, if there is no signal recorded on the tape and providing the tape is completely demagnetized, the tape will be completely noise free.

In determining the signal-to-noise ratio of the overall system, the maximum signal from the playback amplifier, generally with 3% distortion, is compared to the playback of a completely erased tape. Since the tape when completely erased will give no noise, the only noise present in the system is from the playback amplifier. Therefore, the signal-to-noise ratio of tape recorders is determined by the playback amplifier.

NOISE-REDUCTION TECHNIQUES

Great care is taken in tape recorder design to reduce the noise inherent in the input stage of the amplifier. The reason for particular care is that when the signal has been amplified to a sufficiently high level in the first stage, noise contributed by the second and third stages of the amplifier is relatively insignificant.

It is interesting to note that the more expensive the tape recorder, the more care given to the reduction of noise in the first stages of the playback amplifiers. The technique most commonly used is to employ very low-noise tubes or transistors. The 6SJ7, the 6J7, and the 5879 are all well-known, time-tested friends of the experienced audio engineer. These tubes, all pentodes, possess high gain with low noise. A typical circuit using an ECL82/6BM8 tube as a combined af amp and bias oscillator is shown in Fig. 8-2.

In the field of triodes, the 12AY7, a very low-noise tube, was developed. Several home recorders use the 12AX7, which gives very

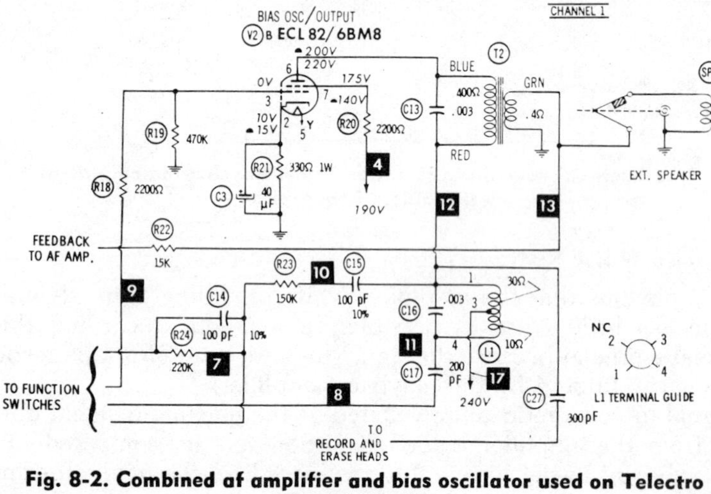

Fig. 8-2. Combined af amplifier and bias oscillator used on Telectro Model SR-122.

high gain but has a tendency to be somewhat noisy. The 6EU7 is a commonly used twin triode and has excellent hum and noise characteristics (Fig. 8-3).

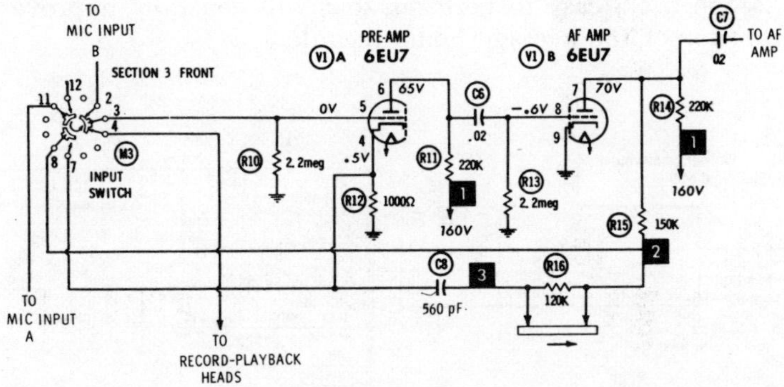

Fig. 8-3. VM Model 740-1 input circuit using a 6EU7 tube.

TYPES OF NOISE

The most common type of noise in a tube is *hum*. Hum is most often introduced from the alternating current used to light the filament of the tubes. In some recorders, as shown in Fig. 8-4, direct current is used to light the filament of the input tube. This, of course, eliminates any possibility of hum being introduced from the tube heaters. However, in other recorders alternating current is used on the filament. In this case it is very important that the filament supply be grounded at a neutral point. The cathode also must be adequately bypassed to eliminate hum (Fig. 8-5). Frequently hum in the input stages is due to the cathode-bypass capacitor losing capacitance.

Another common type of noise in a tube is *shot noise*. This is caused from the electrons leaving the cathodes in groups or clusters rather than in a continuous stream. Because of the irregularity with which these electron groups strike the plate of the tube, a noise similar to the rush of a waterfall is produced. Low-noise tubes were especially designed to reduce this effect.

Still another type of noise common in input stages is *thermal noise*. That is, if a current flows through a resistance a certain type of noise is generated. It is also called *white noise* since it covers a wide band of frequencies at almost constant amplitude.

In designing input stages, the more expensive recorders use wirewound or carbon-deposited resistors rather than the less expensive cast-type ones to eliminate noise. Often the simple replacement of the plate and cathode resistors in a home recorder with wirewound or carbon-deposited resistors will produce a tremendous reduction in the noise in the input stage. While these resistors do not represent much money as a replacement item for a single recorder, cost makes their use prohibitive in mass production of home-type recorders. This is one trick, easy to perform, that will generally improve the performance of the average home recorder.

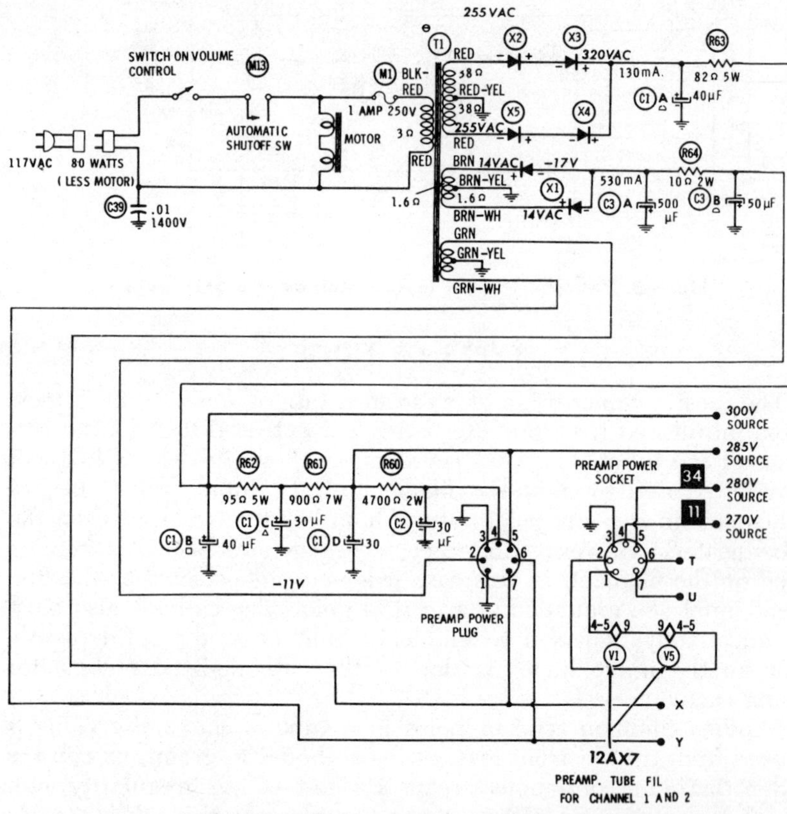

Fig. 8-4. Dc filament supply used in the Webcor Model EP-2207-1.

Another type of noise common in the input stages of amplifiers is a *microphonic* condition. This type of noise can be caused by either microphonic tubes or microphonic parts such as a resistor. Microphonic means that a component is sensitive to vibration and will alter its electrical characteristics when vibrated, operating in a manner somewhat similar to a microphone. The worst microphonic offenders are tubes in the input stage. These can be readily detected by tapping the tube. Generally a loud ringing sound will result.

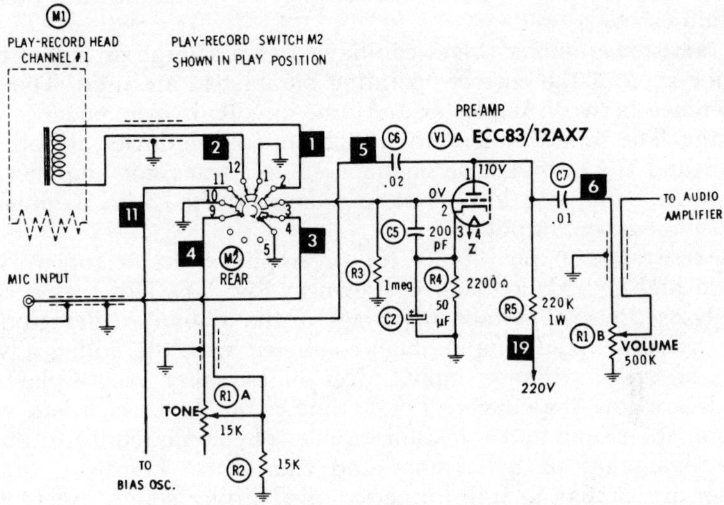

Fig. 8-5. Input stage showing cathode-bypass capacitor and plate circuit of a tape recorder.

Microphonics are especially bothersome in compact, home-type recorders where the amplifier is mounted next to the motorboard. The motor, of course, creates considerable vibration, which is readily picked up by any input tube that is microphonic. Also, since the speaker is mounted in the same case as the amplifier, the vibrations from the speaker will often feed back through the input tube, causing an oscillation or howl to build up when the volume control is advanced.

To eliminate this condition, low-noise tubes were also designed to be as nonmicrophonic as possible. However, it is often necessary to specially select tubes for low microphonics. In some cases, such as in the case of 12AX7 tubes, it is necessary to select perhaps one out of ten or one out of twenty tubes to get a really low-microphonic tube. The input tube of practically all recorders is mounted on rubber mounts that isolate it as much as possible from any vibration. Some manufacturers even hang weights on the tube itself to give it inertia, helping damp out any transmitted vibration. All of these methods help reduce the bothersome microphonic problem.

The technology of transistors has modified the design of amplifiers somewhat compared to tube amplifiers. The standard engineering practices as applied to vacuum tubes are changed or reevaluated to obtain hi-fidelity performance from transistor amplifiers. Transistor circuits are generally no more complicated than tube circuits and sometimes much less so. A transistor does not amplify in the same manner as a tube, but it can be compared to a tube in certain cases.

Transistor amplifier stages can have a gain as high or higher than a tube stage if the correct operating parameters are used. The real difference between transistor and tube circuits is in impedance and biasing. The tube is inherently a high-impedance device at both the input and the output. The common-emitter transistor amplifier, on the other hand, is a low input impedance device with medium or high impedance output.

A resistance-coupled transistor preamplifier circuit for working into a high-impedance load is shown in Fig. 8-6. This was a commonly used circuit to take advantage of the low-noise characteristic of a transistor in the input stage combined with the added advantage of low-impedance input. Most high-quality record-playback heads are low impedance. For a time many tape recorders were combination tube and transistor circuits (hybrid circuits), utilizing the advantages of both tubes and transistors. Transistor output stages are similar to transformer-coupled driver stages except that the emitter and base resistance values are lower because of greater current flow in the output transistor.

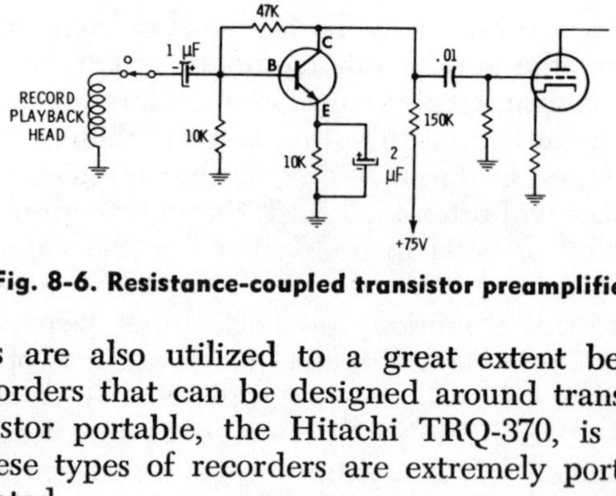

Fig. 8-6. Resistance-coupled transistor preamplifier.

Transistors are also utilized to a great extent because of the compact recorders that can be designed around transistor circuits. An all-transistor portable, the Hitachi TRQ-370, is illustrated in Fig. 8-7. These types of recorders are extremely portable and are battery operated.

HARMONIC DISTORTION

While the primary purpose of both the record and playback amplifiers is to amplify or provide gain, they must give this gain while introducing as little noise and distortion as possible. There are, basically, three types of distortion—harmonic distortion and two types of intermodulation distortion.

Harmonic distortion relates to changes in waveform of any single frequency. Intermodulation distortion relates to changes that occur when two or more frequencies are present at the same time and is thus more directly related to the kinds of distortion that are really noticeable.

In theory, there is a correlation between the various types, but this gets complicated, and some forms of the complex distortion are not directly related to the simpler forms. However, because the concept of distortion can most easily be understood with reference to simple waveforms, this discussion will be based on harmonic distortion.

Harmonic distortion refers to the fact that when a signal frequency such as a pure 400-Hz tone is recorded, not only is the 400-Hz tone received during playback but also an 800-, a 1600-, and a 3200-Hz tone, etc., are reproduced. These are the even-order harmonics or even multiples of the fundamental frequency. Also reproduced from a 400-Hz tone will be a 1200-Hz tone, a 2000-Hz tone, a 2800-Hz tone, etc. These are the odd-order harmonics or odd multiples of the fundamental frequency.

In magnetic recording, it is theoretically impossible to get even- or second-harmonic distortion. If an even-harmonic distortion condition exists, its cause will originate in the amplifier or be due to magnetized tape. One of the outstanding characteristics of magnetic recording that has made it today's quality recording medium is the fact that the process is inherently free of second-harmonic distortion.

Whenever a signal stronger than the tape can record is introduced, thereby overloading or saturating the tape, the tape reaches a point at which it can no longer be magnetized. Therefore, with additional increases in the signal strength there will be no increase in the tape output, causing an effect known as clipping. Clipping, in turn, results in third-harmonic distortion.

The clipping that results from tape overload is not as severe as in other recording methods. For example, in the disc recording process, if the disc is overrecorded, one actually cuts into the adjacent grooves of the recording, resulting in almost total ruin as shown in Fig. 8-8. In photographic recording, the peaks of the waves are chopped flatly off by overrecording, causing a very harsh type of distortion as shown in Fig. 8-9. However, in magnetic recording, overrecording tends to flatten the peaks of the curve in a gradual manner as shown in the graph in Fig. 8-10. This means that although third-harmonic distortion is introduced, it is far more gradual than for other mediums.

Using disc or photographic recording systems, the recording level must be carefully guarded at all times because disastrous results happen if the recording is made at too high an amplitude. In contrast, using the magnetic-recording medium, even if the amplitude

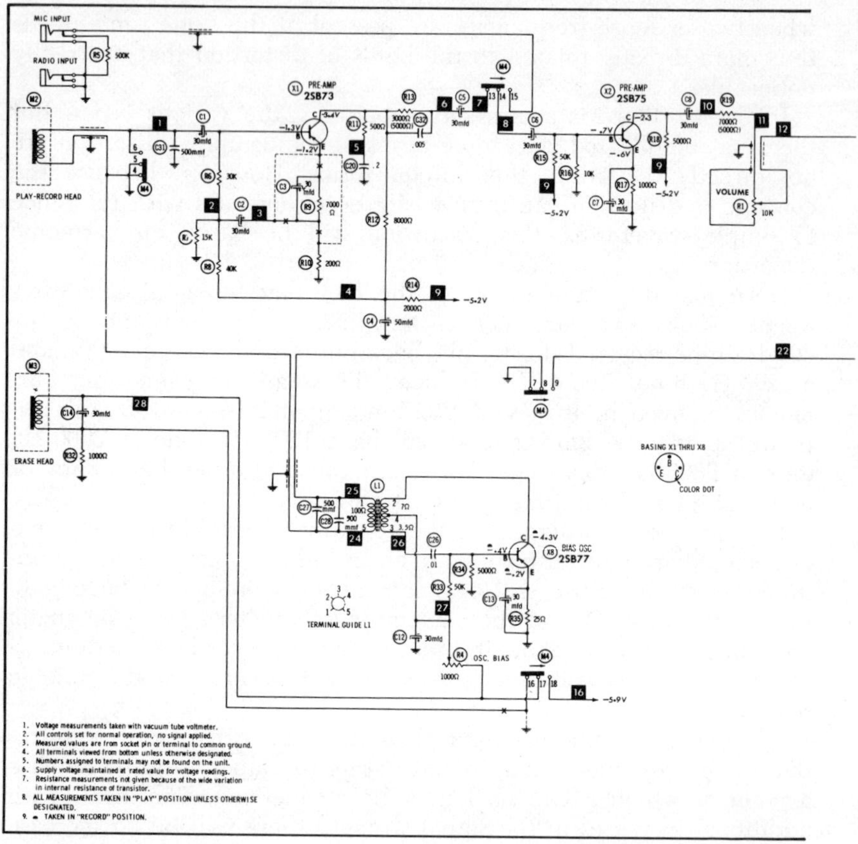

Fig. 8-7. Hitachi TRQ-370

is several times higher than the normal record level, the resulting distortion will often be unobjectionable.

Using a magnetic-tape recorder, it is possible for an inexperienced 12-year-old child to make a totally acceptable recording. Understanding this inherent advantage, it is not difficult to realize why magnetic tape recording has become so popular as a recording medium.

We have seen that second-harmonic distortion is theoretically nonexistent and third-harmonic distortion is introduced only by overloading or overcorrecting the tape. However, if we measure recorders, we may see both second- and third-harmonic distortion in the output amplifiers. Second-harmonic distortion is mainly introduced by faulty design in the amplifier circuits. The better recorders use large amounts of feedback together with push-pull out-

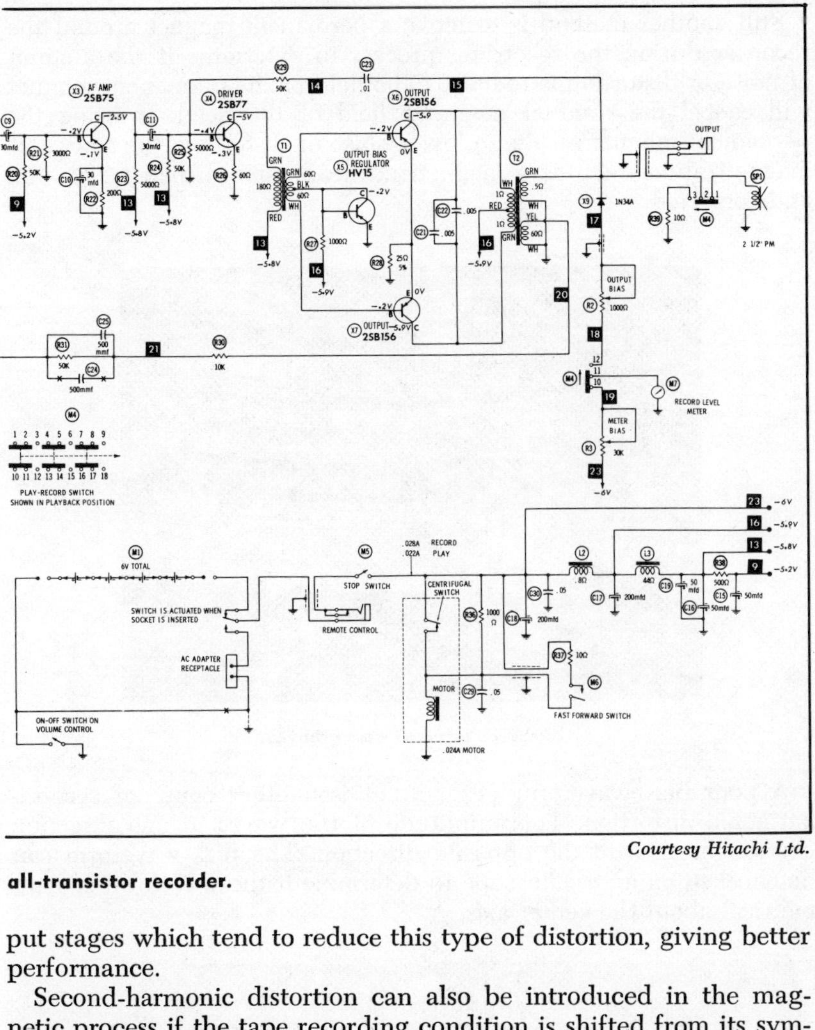

Courtesy Hitachi Ltd.

all-transistor recorder.

put stages which tend to reduce this type of distortion, giving better performance.

Second-harmonic distortion can also be introduced in the magnetic process if the tape recording condition is shifted from its symmetrical counterpoint. That is, if the tape is at all magnetized during the recording process, severe second-harmonic distortion is possible. The magnetization of the tape can be caused by any one of the magnetic heads becoming magnetized or from a faulty erase system.

The preferred method of checking for magnetized heads is first to demagnetize the heads with a head demagnetizer to determine if the performance has improved. If the diagnosis was correct, there will be a drastic reduction in noise followed by a reduction in distortion.

Still another method is to move a permanent magnet around the recorder during the recording process to determine if the amount of noise or distortion is reduced. The field of the permanent magnet will cancel the residual magnetic field of the head. If using the permanent magnet results in lower noise or distortion, the source of magnetization, such as a magnetized head, should then be located and corrected.

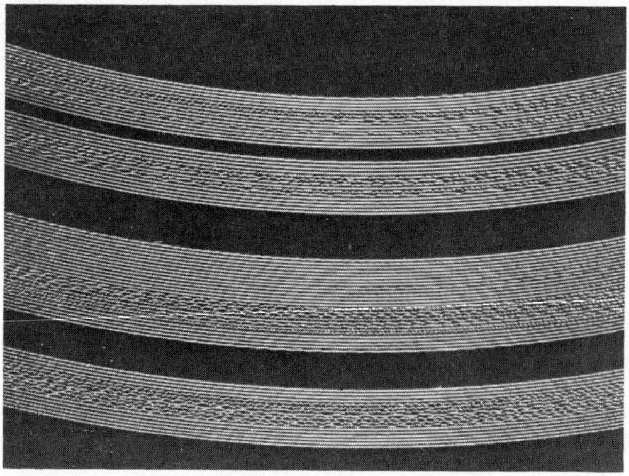

Fig. 8-8. Overcut disc grooves.

A poor bias waveform (Fig. 8-11) is another cause of second-harmonic distortion. The amplitude of the waves in one direction is stronger than in the opposite direction. The bias waveform can be checked on an oscilloscope to determine if the waveform is symmetrical about the center axis.

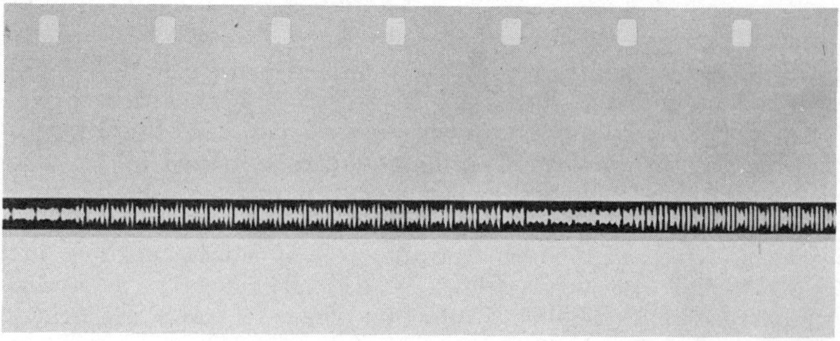

Courtesy 3M Company

Fig. 8-9. Overmodulated optical sound track.

The distortion introduced in the amplifier itself is sometimes caused by faulty original design, such as a pentode output stage with insufficient feedback. Other possibilities are improper bias on the tubes or transistor caused by a resistor changing value or leaky coupling capacitors resulting in a positive voltage changing the normally negative bias voltage.

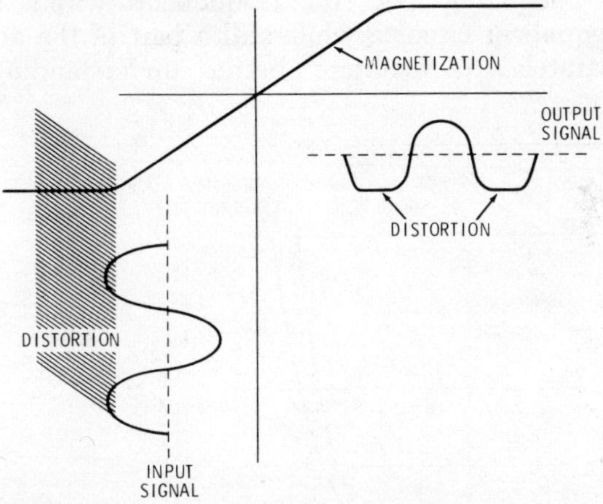

Fig. 8-10. Distortion caused by partially magnetized tape.

As discussed in the previous chapter, bias level can affect the amount of and the way third-harmonic distortion sets in as maximum level is approached. Fig. 8-12 shows how this varied for one of the earlier high quality ferrous oxide tapes. Transition to chro-

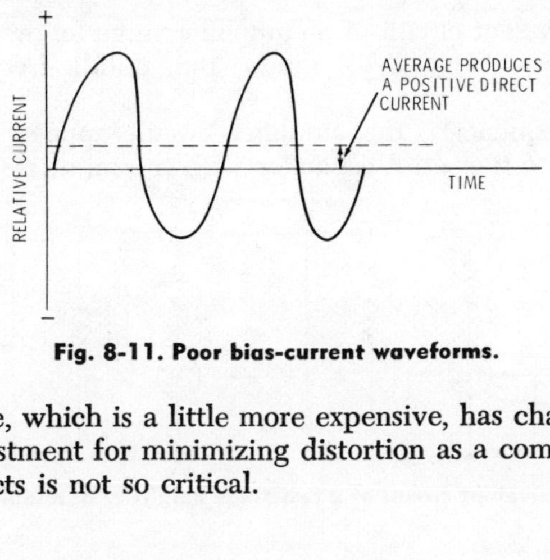

Fig. 8-11. Poor bias-current waveforms.

mium dioxide, which is a little more expensive, has changed this, so that the adjustment for minimizing distortion as a compromise with its other effects is not so critical.

FREQUENCY RESPONSE

In addition to amplification gain, low noise and distortion, an amplifier must also have good frequency response. In the chapter devoted to equalization, it was explained that the type of equalization chosen primarily determines the overall frequency response of the recorder. However, the amplifier circuits themselves must be designed to adequately pass the frequencies within the desired range. The equalizer circuits, while still a part of the amplifier, are treated separately to facilitate better understanding of their operation.

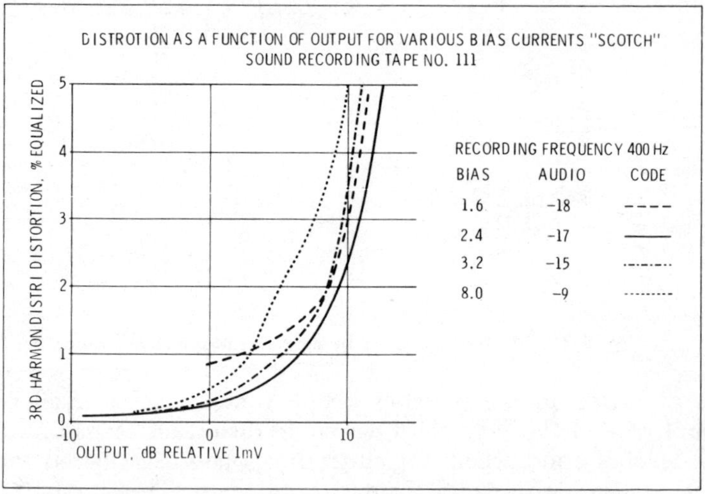

Fig. 8-12. Distortion as a function of signal level.

In an equivalent circuit of an amplifier stage followed by another amplifier stage, Fig. 8-13 shows the middle-frequencies configuration.

At low frequencies, the amplifier circuit appears as shown in Fig. 8-14. Note the series capacitor with the input resistance of the

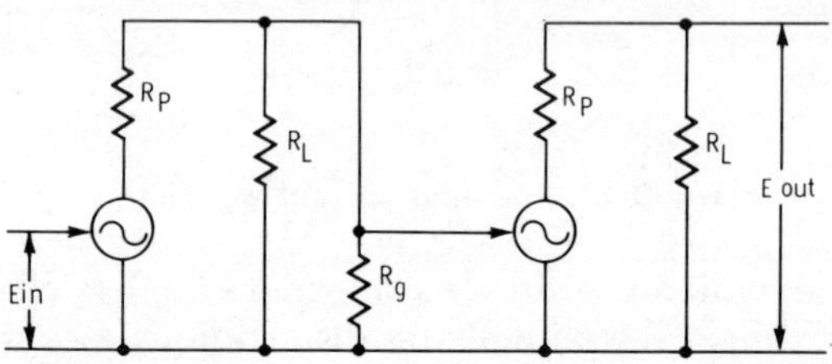

Fig. 8-13. Equivalent circuit of a two-stage amplifier at middle frequencies.

following stage. This capacitor at low frequencies has a high reactance and serves to pass high frequencies and reject low frequencies.

To improve the low-frequency response of an amplifier, the size of the coupling capacitors, therefore, should be increased. The larger the capacitor, however, the higher the cost. Because of cost considerations, most manufacturers generally use as small a capacitor as possible, settling for the minimum acceptable low-frequency

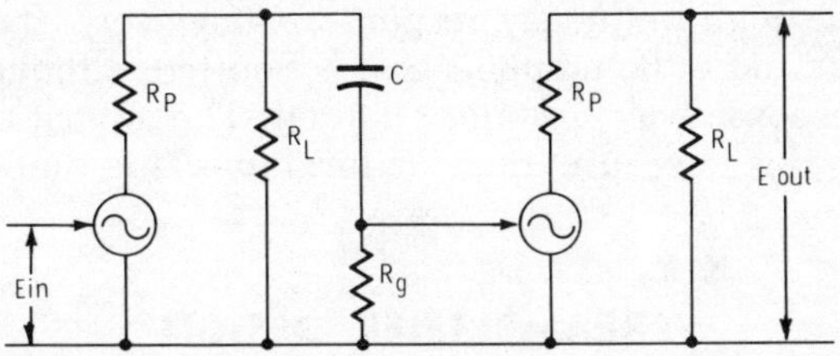

Fig. 8-14. Equivalent circuit of a two-stage amplifier at low frequencies.

response. Also, particularly in transistor circuits, a capacitor that is larger than necessary can either promote instability, or increase the time the amplifier takes to recover from instability caused by excessive signal (program) peaks. It should also be remembered that there are other factors besides inefficient capacitors that limit the low-frequency response. These factors are covered in the chapters that discuss equalization and magnetic heads.

The high-frequency equivalent circuit appears in Fig. 8-15. Note that the equivalent input capacitance of the following stage then serves to limit the high frequencies. In other words, at low frequencies the bypassing effect of the capacitance is small compared to the input resistance. So this equivalent capacitance has no effect on the circuit. However, at higher frequencies this capacitance tends to short out the incoming signal and limit the high-frequency response. The equivalent capacitance of the following stage cannot be changed except by changing circuit design, which generally

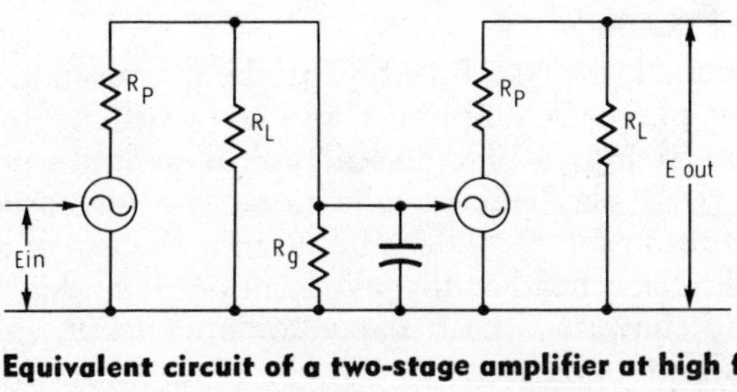

Fig. 8-15. Equivalent circuit of a two-stage amplifier at high frequencies.

does not substantially improve the situation. Therefore, it is necessary to lower the load resistance of the preceding stage. This, in effect, makes the reactance of the shunt equivalent capacitance small in respect to the load resistance of the preceding stages. While this design will pass higher frequencies, the gain will be adversely affected.

Therefore, in a tube amplifier circuit, the value of load resistance in the circuit is a compromise between gain or amplification and high-frequency response. The lower the resistor, the less gain and the better the high-frequency response. Conversely, the higher the resistance, the more the amplification is boosted, although the high-frequency response becomes more limited. The advent of transistors changed all this, because their natural working impedances are much lower.

PREAMPLIFIER DESIGN

Due to the large amount of design experience gained in vacuum-tube circuits and also to the large number of inexpensive, low-noise vacuum tubes that are available to the designer the vacuum tube held a predominant position in tape-recorder amplification. Now the reduced cost and size of transistors, coupled with their improved performance, has made them the preferred choice of all modern designers.

The tape-recorder preamplifier must be designed with the following points in mind:

1. Signal-to-noise ratio must be as high as possible; on professional equipment, tape noise should be the limiting factor in the amplifier.
2. Equalization should permit flat system response with a tape recorder to NAB or similar standards.
3. The preamplifier should be stable, and, if possible, feedback should be used to reduce gain variations due to component aging.

The Pentode Preamplifier

The high-gain, low-noise linear pentode was used extensively in tape-recording playback amplifiers and may still be found in use. The schematic in Fig. 8-16 shows a typical pentode unit designed for use with NAB playback equalization. The increased input impedance inherent in the pentode tube permits the use of a very-high–inductance playback head with degeneration at higher frequencies. The schematic shows a .01-μF capacitor and a 20K potentiometer in a simple shunt equalizer. The value of the capacitor must be

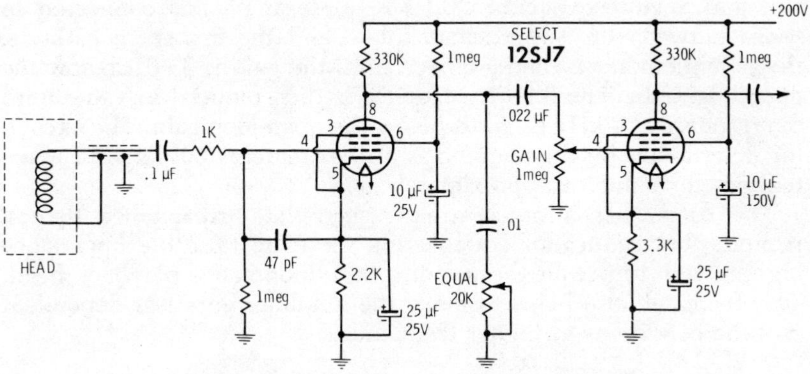

Fig. 8-16. Typical pentode preamplifier circuit.

chosen so that the absolute value of its capacitive reactance is equal to the effective pentode output impedance at a frequency between 20 and 60 hertz; the actual frequency depends on the low-frequency characteristics of the playback head being used. The value of the potentiometer is determined by the high-frequency performance of the head and the tape speed being used.

The Triode Preamplifier

The amplifier shown in Fig. 8-17 is a triode unit with the equalization applied in the feedback circuit from the plate of the second stage to the cathode of the first stage. A simple method of determining the value of the feedback parameters is as follows: First, the gain of the two-stage amplifier without feedback is determined at a frequency of 1 kHz. A network consisting of the 500K potentiom-

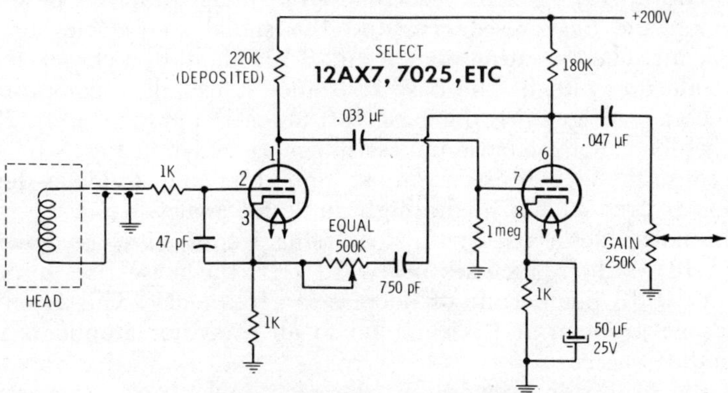

Fig. 8-17. Triode amplifier circuit.

eter and a large capacitor (0.1 μF, for example) is connected in series between the second-stage plate and the first-stage cathode; the potentiometer is then adjusted until the gain is 35 dB below the open-loop gain. The 0.1-μF capacitor is then reduced in value until the response at 1 kHz is 25 dB below the open-loop gain. The capacitor determined by this method is approximately 750 pF with a potentiometer value of approximately 50K.

The triode has a much greater grid-plate capacitance than a pentode, but the use of equalization serves to raise the impedance high enough to permit the use of a high-inductance playback head. This feedback also helps to make the amplifier gain less dependent on tube conditions at higher frequencies.

Transistor Playback Preamplifier

The transistor is rapidly becoming universal in new equipment design. In inexpensive home machines, current transistor prices permit manufacturers to compete with tube designs, since filament supplies, tube sockets, and associated components are eliminated. In quality equipment, more reliable transistors cost more. But transistors still have the edge on tubes because their long-term reliability is much better than tubes, which have a limited life before they need replacing.

A typical circuit (Fig. 8-18) is stabilized to prevent variations in performance with high operating temperature. This is accomplished by several techniques, including negative feedback from the emitter of the second stage to the base of the first stage and direct coupling from the first-stage collector to the second-stage base. Both stages also have relatively large bypassed emitter resistors.

Obtaining a good signal-to-noise ratio involves a number of factors. The operating point (current and voltage) of the base and collector, the base resistance, and the signal frequencies are all significant. The operating point (Fig. 8-19) should be chosen to use low collector voltage. The base resistance is usually a compromise. As shown in Fig. 8-18, the base resistance is approximately 20 to 25K, which is not the optimum resistance as shown in Fig. 8-19. This is because the base resistance must be high enough to avoid degeneration of frequencies in the high audio-frequency band.

The noise factor rises with decreasing frequencies, as shown in Fig. 8-19. With frequencies below 10 kHz, the noise rises approximately 10 dB per decade of decrease in frequency. This is an unavoidable problem and is common to all transistor amplifiers used for audio.

In the transistor preamplifier the record/playback head impedance should be low enough over the entire audio band of frequen-

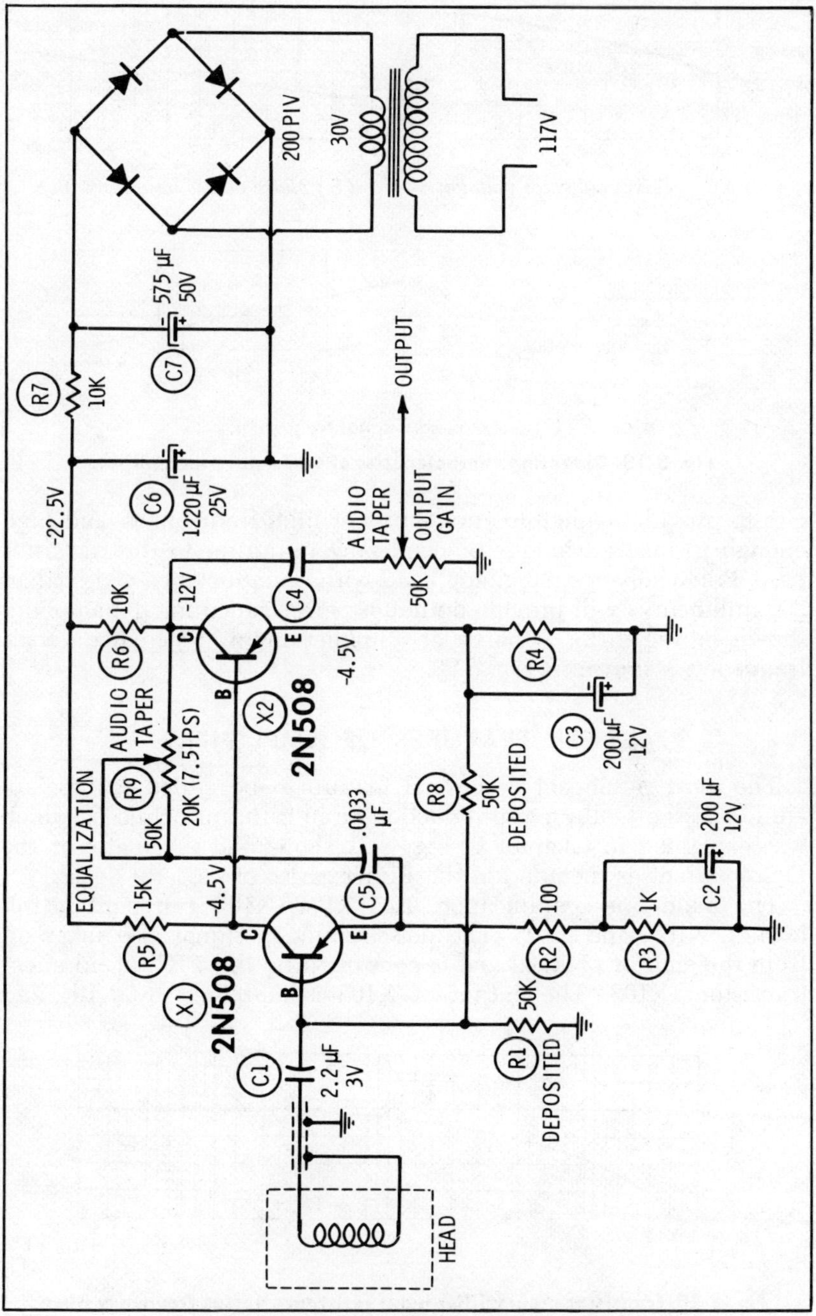

Fig. 8-18. Transistor playback preamplifier.

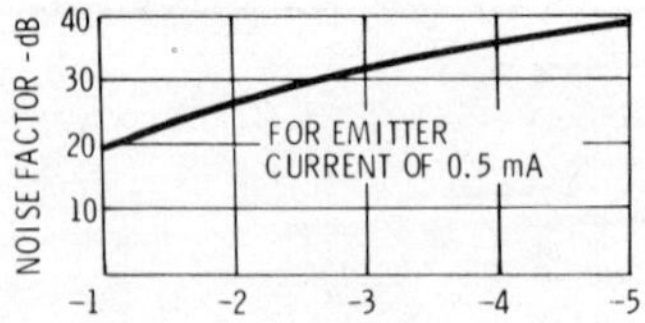

(*A*) *Noise versus collector voltage.*

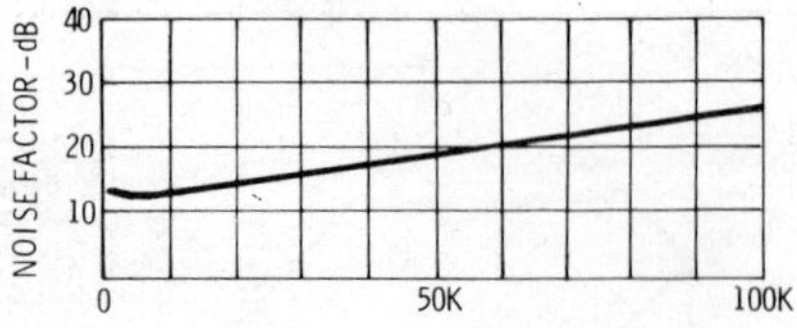

(*B*) *Noise versus base resistance.*

NOISE FACTOR IN dB

$$= 10 \log_{10} \frac{S_{IN}/N_{IN} \text{ (POWER)}}{S_{OUT}/N_{OUT} \text{ (POWER)}}$$

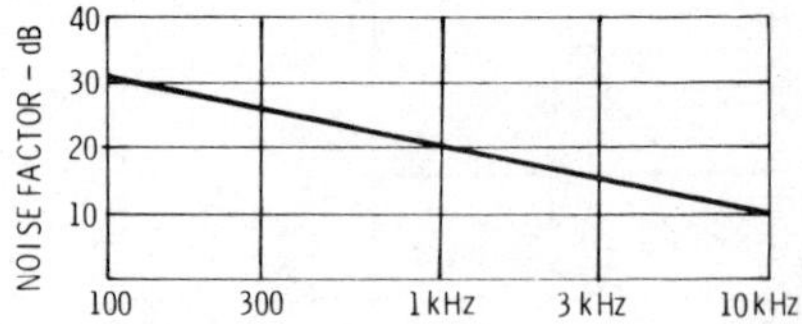

(*C*) *Noise versus signal frequency.*

Fig. 8-19. Operating characteristics of transistor preamplifier.

cies to provide negligible attenuation of high frequencies, and high enough to take advantage of good power transfer to the transistor base. Practically speaking, any head with an inductance of less than 200 millihenrys will provide optimum performance for the amplifier shown in Fig. 8-18. A curve of amplifier input impedance versus frequency is shown in Fig. 8-20.

OTHER FEATURES OF AMPLIFIERS

The most significant additional feature to be added to tape recorders is the Dolby noise reduction circuit, the principle of which was explained in Chapter 6. Fig. 8-21 shows the schematic of the Dolby circuit as included in the Fisher series of cassette decks.

The main line amplification uses X101, X102 (an emitter follower), X107, and X108. The noise-reduction signal also takes off from the emitter of X102 and is controlled by the FET (field effect transistor) X103. The output of X103 is amplified by X104 and

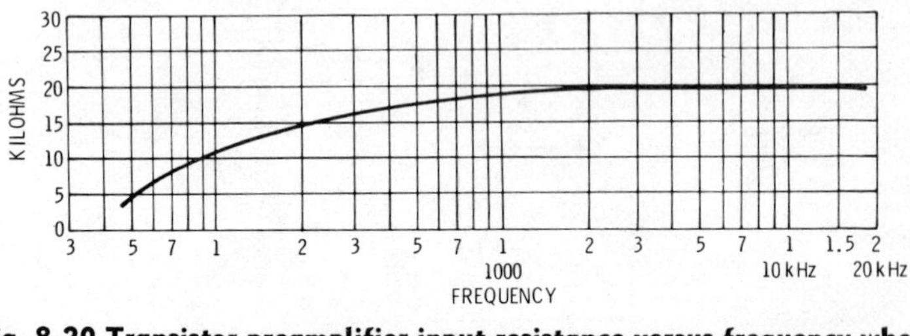

Fig. 8-20. Transistor preamplifier input resistance versus frequency when equalization is set for NAB curve.

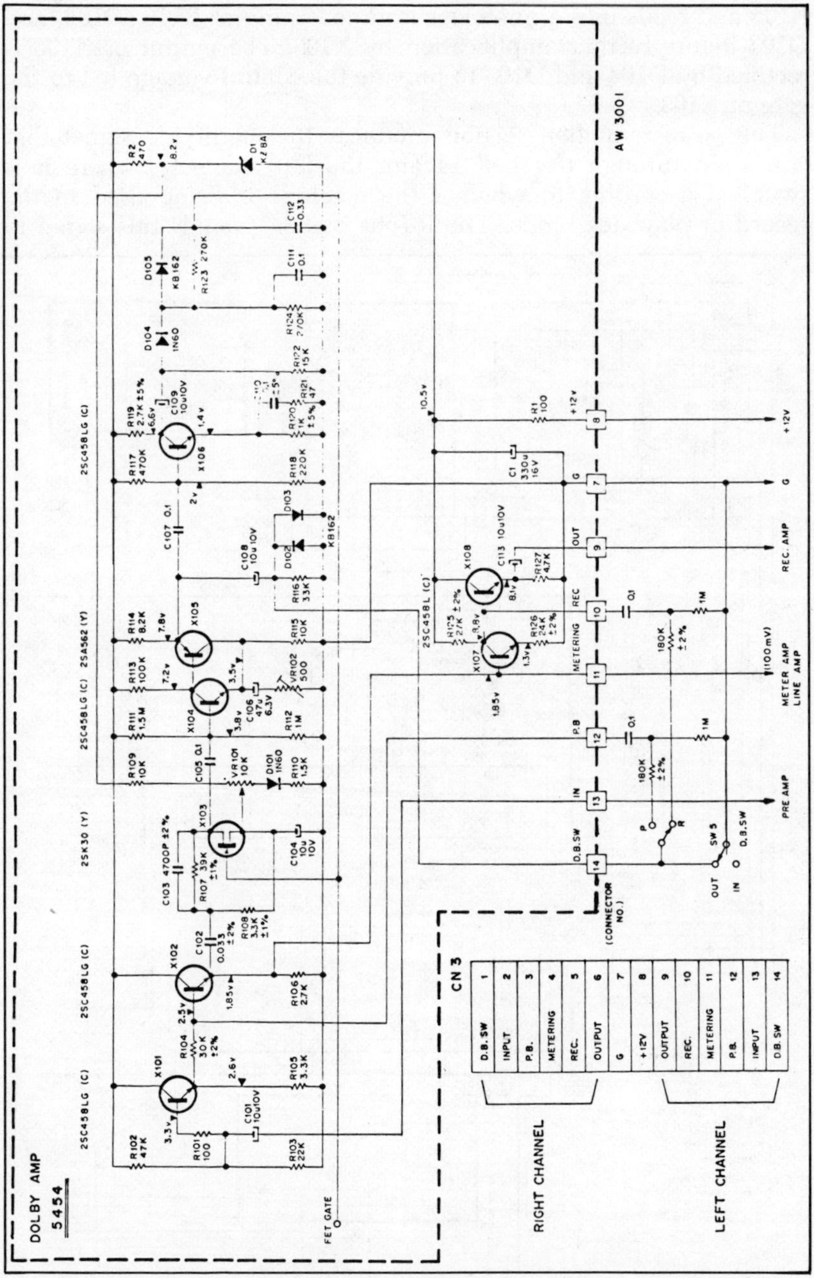

Fig. 8-21. The complete Dolby circuit for one channel as used in Fisher cassette recorder decks.

X105 and feeds into a nonlinear lead consisting of R116, D102, and D103 before further amplification by X106. The output of X106 is rectified by D104 and D105 to provide the control voltage fed to the gate of X103.

The noise reduction section produces the modifying signal that is fed out through terminal 14 (for the left channel), where it is switched according to whether the machine is being used in the record or playback mode. The in/out switch grounds this signal to

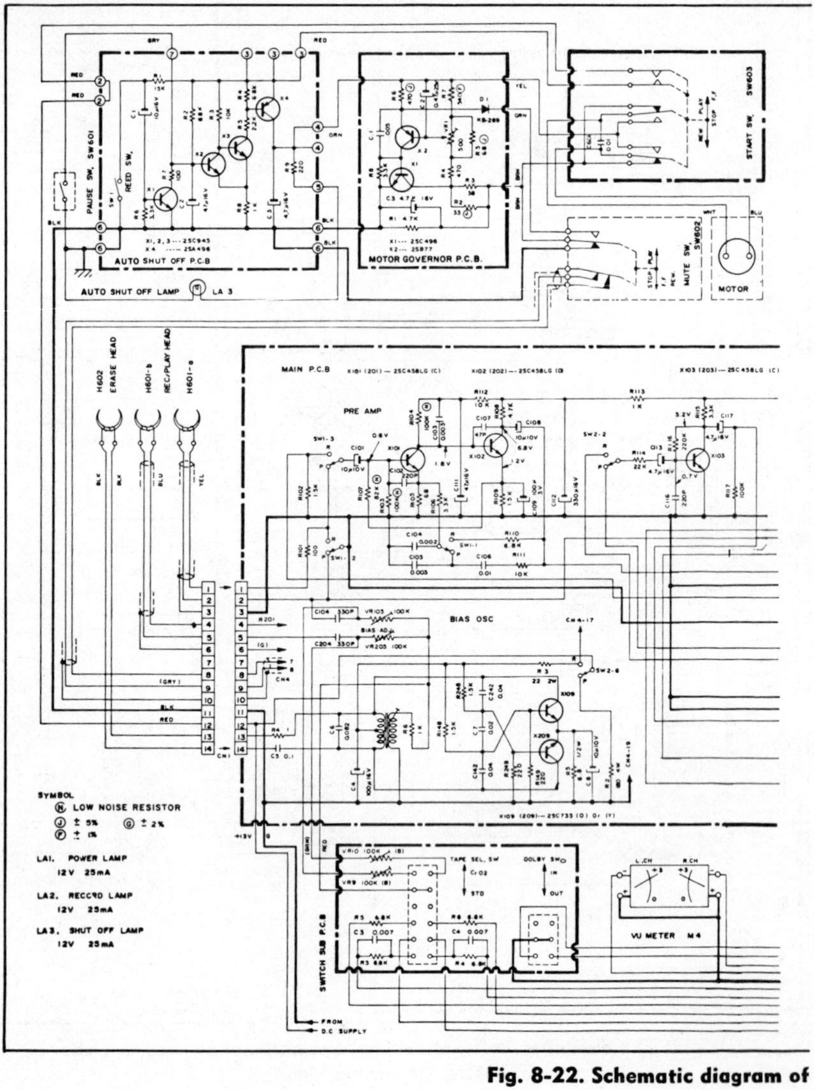

Fig. 8-22. Schematic diagram of

disable the noise-reduction action, so only the main line amplification is operative.

The record/playback switch connects the remix point forward to the base of X108 for record and backward to the base of X102 for playback. Phasing is taken care of, so that on record the extra signal augments weak signals and does very little to full-level signals. On playback, the weak signals that were built up are reduced by negative feedback, while the original stronger signals are not affected.

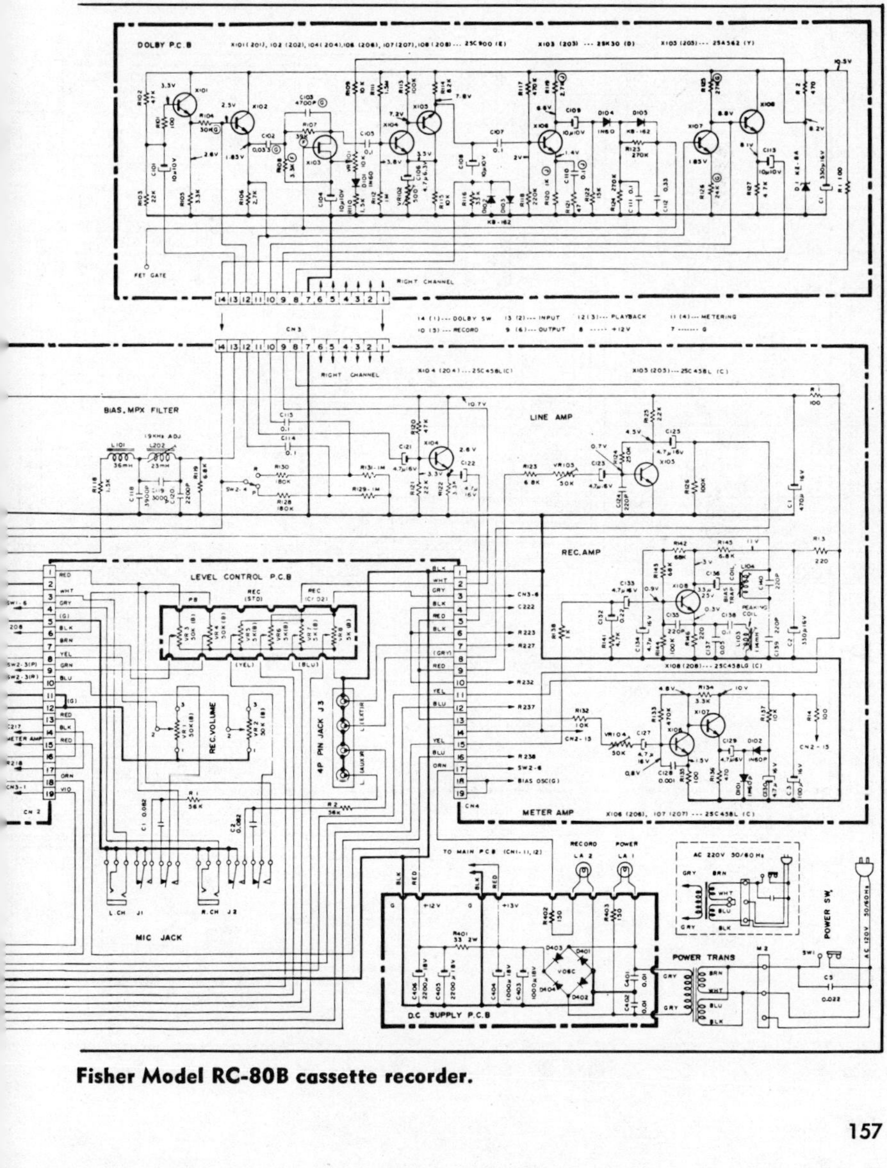

Fisher Model RC-80B cassette recorder.

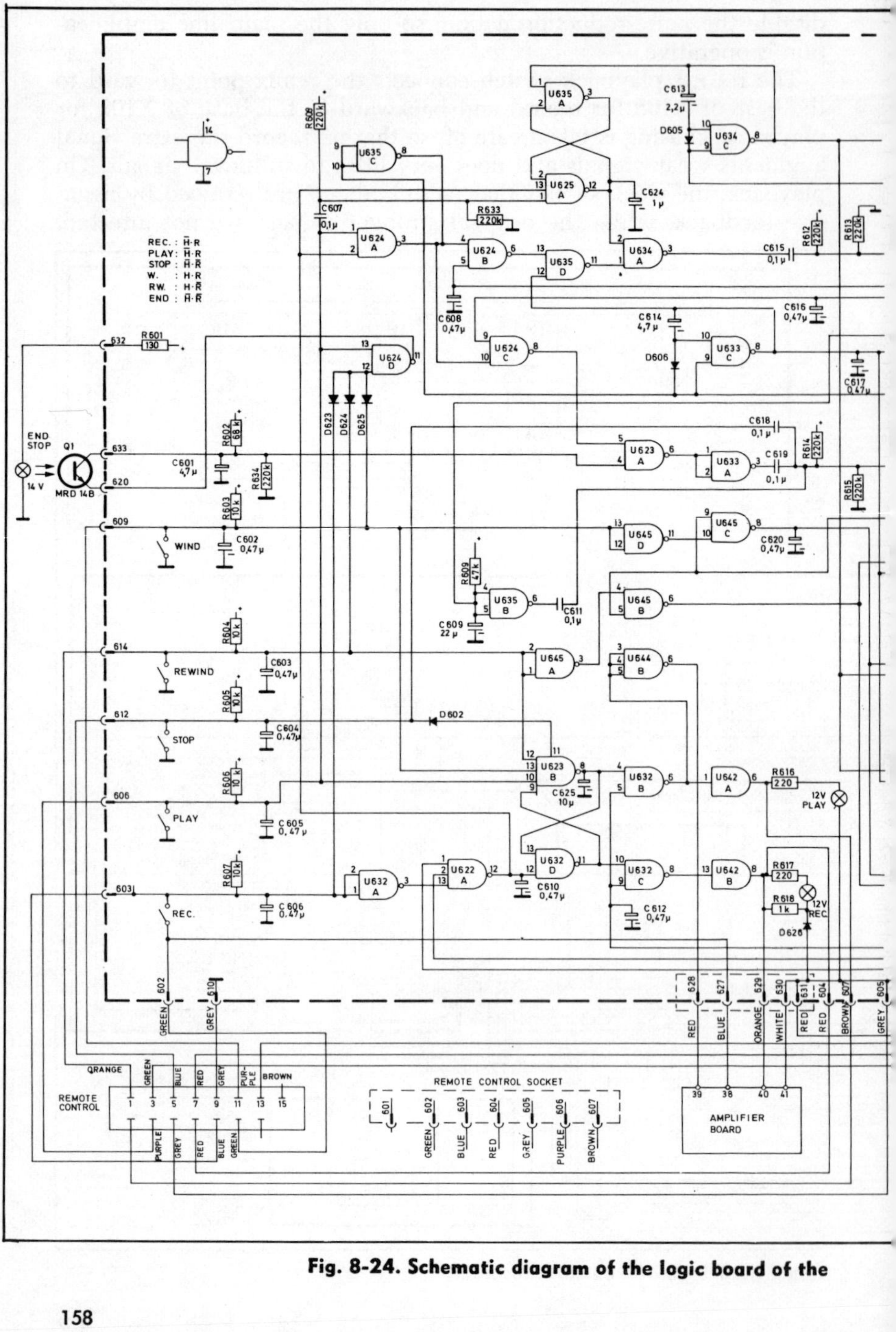

Fig. 8-24. Schematic diagram of the logic board of the

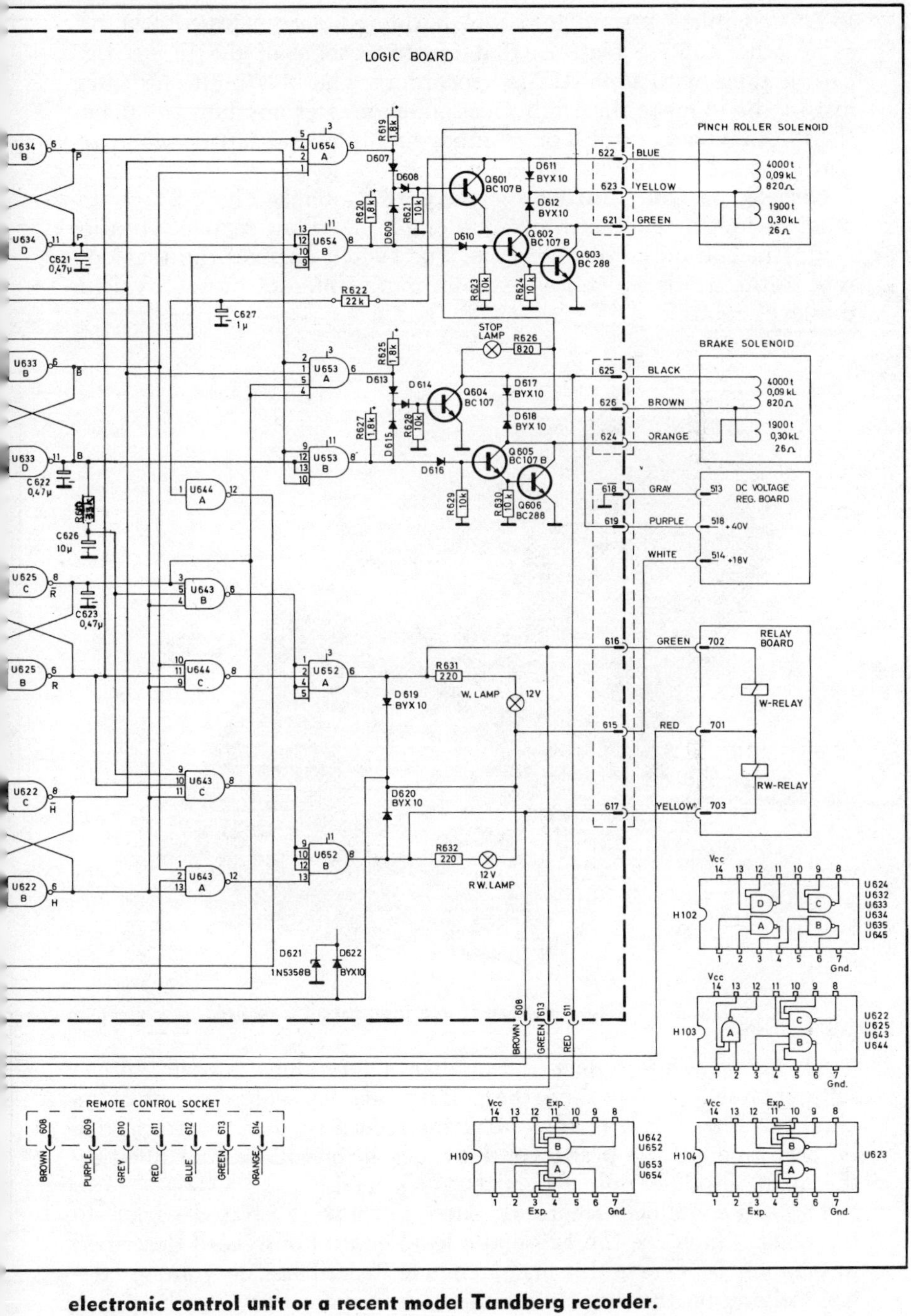

electronic control unit or a recent model Tandberg recorder.

The complete system into which this connects (Fig. 8-22) has some other unique features that illustrate some of the things that can be done with sophisticated recorders. The bias/multiplex filter, just by the input to the Dolby section, eliminates any bias frequency (in the playback mode) or residual multiplex subcarrier when recording stereo multiplex program.

Next to the Dolby in/out switch on this model (Fig. 8-23) is another switch that alters the recorder for handling regular (ferrous oxide) or chromium dioxide tape. The switch changes the bias current to the heads as well as establishing a different preset level for the level control.

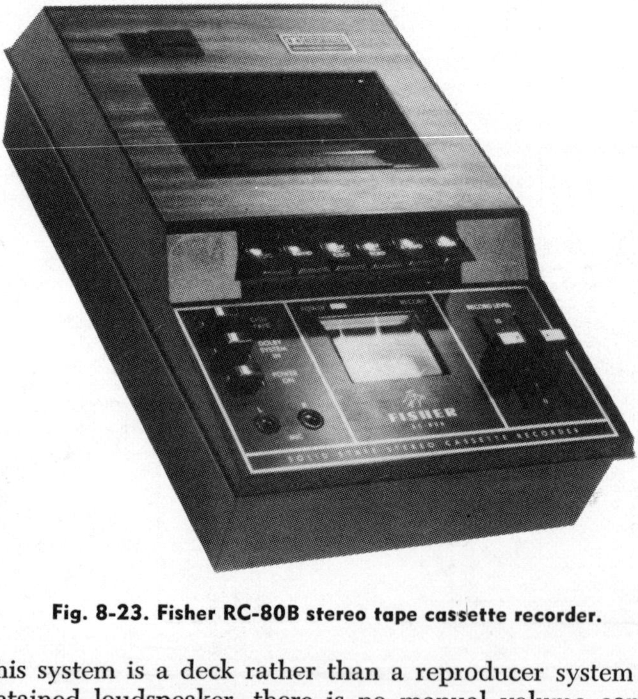

Fig. 8-23. Fisher RC-80B stereo tape cassette recorder.

As this system is a deck rather than a reproducer system with a self-contained loudspeaker, there is no manual volume control on playback. There is a level control for record so that recording can be performed at the best level. Playback level will be controlled by the main amplifiers into which the deck feeds.

The self-contained recorders, such as those illustrated earlier in the book, which use the automatic level control on record described in Chapter 6, reverse this arrangement. Record has no volume control, relying on the automatic control to handle it, while there is a control on playback that allows you to adjust the volume. Each arrangement suits its own purpose best.

The motor governor used in this system controls voltage precisely, with a somewhat different circuit from that shown in an earlier chapter. The automatic shutoff feature also utilizes the same circuit to cut off power to the motor.

We are living in days when technology is changing rapidly. The latest Tandberg recorders have logic circuits to control the entire tape handling set of operations. Fig. 8-24 shows the schematic of the logic board of this recorder. It uses a photoelectric end stop to avoid undue stress on the tape. The logic takes into account how fast the tape is moving when the button is hit, how much of the tape is on each spool of the cassette, and so forth, so it is always handled perfectly.

CHAPTER 9

Magnetic-Recording Heads

The magnetic-recording head is the most critical precision device in the entire recording mechanism. Built to a tolerance as close as 1 millionth of an inch, the magnetic head requires an exceptionally high degree of precision in its construction.

Good fidelity in record operation, at the present-day speeds of 1⅞ and 3¾ ips, has been achieved. This, in large measure, has been made possible through basic improvements in magnetic heads. Magnetic heads not only determine the low- and high-frequency response of a recorder but also help to establish the signal-to-noise ratio.

Some two-head recorders have a combination erase and record/playback head mounted in a single case, giving the appearance of only one head, although these are actually two separately operating heads. Also, some monaural tape recorders use a permanent-magnet erase head which is mechanically moved into contact with the tape during recording.

It is imperative to get the strongest signal possible out of the magnetic head and onto the tape. This is because a strong signal means that less amplification is required to reproduce sound at an audible level. Noise in modern recordings is partly from the tape and partly from the playback amplifier. Consequently, a head that gives a strong signal means that less playback amplification is needed and less noise will be introduced. The magnetic head is, therefore, the very heart of the magnetic-recording process. Fig. 9-1 shows a modern 2-track stereo magnetic-recording head.

In most discussions it is common to speak of three magnetic heads. First, the erase head, which has the function of obliterating or "wiping" off any signal on the tape, leaving the tape in a completely demagnetized condition. Second, the record head which serves to place a signal in the form of magnetic impulses on the tape. And third, the reproduce or playback head which has the function of converting the stored magnetic impulses on the tape into electrical energy to be amplified for reproduction.

Courtesy The Nortronics Company, Inc.

Fig. 9-1. Modern two-track stereo recording head.

In respect to design, the record and reproduce heads are practically identical. In the majority of machines, especially home-type tape recorders, the same head is used for both record and playback functions. This is accomplished by merely switching the head from the output of the amplifier during the recording process to the input of the amplifier during the playback process. Therefore, only the erase head and the record-reproduce head will be discussed.

THE ERASE HEAD

As previously mentioned, the first function of the erase head is to completely obliterate any signal on the tape from the prior recording. The second function is to leave the tape in a completely neutral or demagnetized condition. To completely wipe off any previous signal from the tape is, of course, important since an echo or background of the previous recording would disturb the new recording.

The tape must also be left in a completely neutral or demagnetized state in order to eliminate all distortion and noise.

To accomplish both objectives in tape erasure, a magnetic field which is considerably stronger than the strongest signal on the tape is needed. This field must be designed in such a manner that it will highly magnetize the tape in one direction, obliterating the previous signal and then demagnetize the tape, leaving it in a completely neutral state.

Permanent-Magnet Erase

Many early recorders and a few present-day machines use what is known as permanent-magnet erase. This is a permanent magnet generally made of *Alnico* material constructed so that one pole completely magnetizes the tape, saturating it to a higher magnetic state than the strongest previous signal. Thus, the permanent-magnet erase serves to obliterate any signal on the tape. Then either a second magnet, a series of magnets, or a diagonal magnetic gap is used to demagnetize the tape and leave it in a neutral state. A typical permanent-magnet erase head is shown in Fig. 9-2.

The chief disadvantage of the permanent-magnet erase is that it must be mechanically operated. This is accomplished during playback by moving the erase head away from the tape by a mechanical

Courtesy Arvin Industries, Inc.

Fig. 9-2. Typical full-track permanent-magnet erase head.

linkage so as not to erase the tape. However, the head must be engaged during the recording process. This is somewhat more complicated than the electromagnetic erase head in which only an electrical circuit is opened and closed.

Another disadvantage of the permanent-magnet erase is that it is difficult to obtain a demagnetized tape that is in a completely neutral state. Extreme care must be taken in adjusting the magnets so as to remove the saturated field. One system that involves a diagonal magnetic gap does a very effective job of neutralizing the magnetic field of the tape.

The permanent-magnet erase head, however, has several positive advantages. No strong high-frequency signal source is required, thus eliminating the major demand on the bias supply (see Chapter 7). This simplifies the design of the bias oscillator, meaning that the electronics can be built with a smaller power transformer. Since there is a smaller tube in the bias oscillator, many economies result in the amplifier design. This is far less of an advantage in solid-state (transistorized) machines.

High-Frequency Erase

The high-frequency erase head is a head with a wide gap that uses a very high frequency, generally the bias frequency, to erase the tape. The source for the current in this case is described in the chapter devoted to the bias oscillator. In the high-frequency erase process, as the tape passes over a wide gap (Fig. 9-3A), a series of reversals takes place in the magnetic field on the tape. When the tape leaves the gap, it is in an essentially neutral or demagnetized condition.

Some erase heads are constructed exactly as a recording head except for a wider gap (about .004 inch or 4 mils) and an increased track width. Also, many are of "double-gap" construction because a more perfect erasure can be obtained if the tape passes across two gaps as shown in Fig. 9-3B.

The high-frequency erase head makes an ideal method of tape erasure. If the signal is strong enough and if the head is correctly designed, the high-frequency erase method will obliterate any previous signal, leaving the tape in a completely neutral condition. It should be pointed out that if this head should become magnetized, which will result from suddenly turning the bias oscillator off, the tape will be left in a polarized condition, introducing distortion and noise. Therefore, the erase head should be occasionally checked to be sure that it is demagnetized. A head demagnetizer is a useful tool for all magnetic recorder heads. It is even more important that the record/playback head not be magnetized; and it is more likely to become so because of the smaller gap in its magnetic

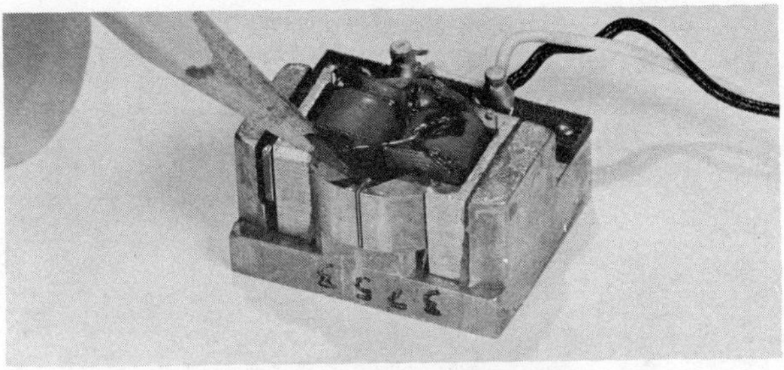

(*A*) *Long-gap erase head.*

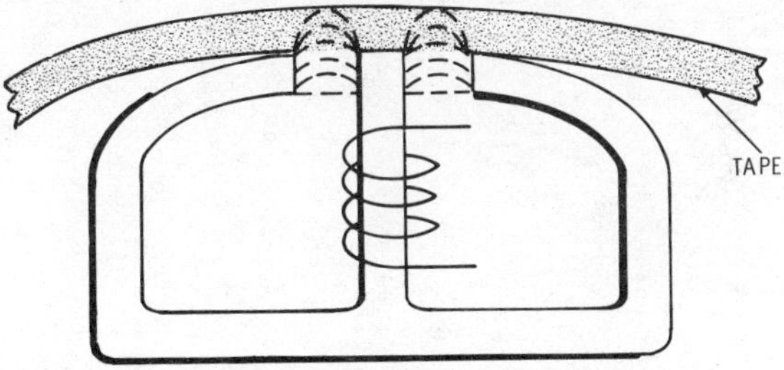

(*B*) *Double-gap erase head.*

Fig. 9-3. Construction of a high-frequency erase head.

circuit. The demagnetizer (Fig. 9-4) should be run slowly across the gap and gradually removed with the current still on. This effectively removes any residual magnetism in the magnetic head.

Bulk Eraser

The final method of tape erasure is accomplished with a bulk tape eraser such as shown in Fig. 9-5.

If a recorder erase system is working right, it will completely erase any normal signal on present-day tapes. However, if a tape is severely overrecorded, or if an old black-oxide tape happens to be used, signals can be completely removed by use of a bulk eraser. For this reason and for the ease with which an entire reel of tape may be completely and cleanly erased, a bulk eraser is used.

The tape is placed on a bulk eraser from which a large alternating-current field, generally 60 Hz, radiates. The tape is slowly rotated in this field and then is gradually and uniformly withdrawn from the field. It is this gradual withdrawal which makes for a perfect bulk erasing job. It is obvious that if the bulk eraser were turned off while the tape was in its vicinity, there would be a good chance of leaving the tape highly magnetized, which is undesirable. Therefore, the tape should be placed on the bulk eraser, the bulk eraser turned on, and the reel very slowly rotated to completely cycle all sections of the tape as shown in Fig. 9-6A. Then, as the tape is gradually withdrawn (Fig. 9-6B), the tape will be first strongly magnetized in one direction and then in the other direction less strongly, then in the original direction still less strongly, eventually wiping off all magnetization.

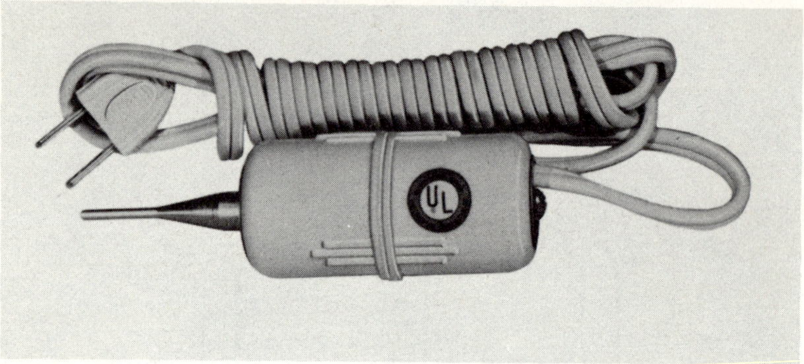

Courtesy Robins Industries Corp.

Fig. 9-4. Head demagnetizer (degausser) to remove residual magnetism from a recorder head.

The two chief errors in bulk tape erasure are that either the tape is rotated and withdrawn from the bulk eraser too quickly or the bulk eraser is turned off with the tape in its proximity. The tape should be withdrawn at least two feet before turning off the bulk eraser to assure complete erasure of the tape. There is no need to erase virgin magnetic tape which is supplied from the manufacturer.

THE RECORD-REPRODUCE HEAD

Since the record-reproduce head is generally the same head in most machines, serving a dual function, this discussion will cover the general design of only one head. Where separate heads are used to achieve the ultimate in overall performance or so playback can monitor what was recorded an instant earlier, the record head uses a wider gap than the playback head but not as wide as the erase head.

In general, the magnetic record-reproduce head is a closed magnetic circuit with a gap at one point. The physical configuration is the reason why it is sometimes called a ring head (Fig. 9-7). On the ring is wound a coil to which the electrical signal is applied during record, and during playback the signal is drawn from the same coil.

(A) Conventional home-type eraser.

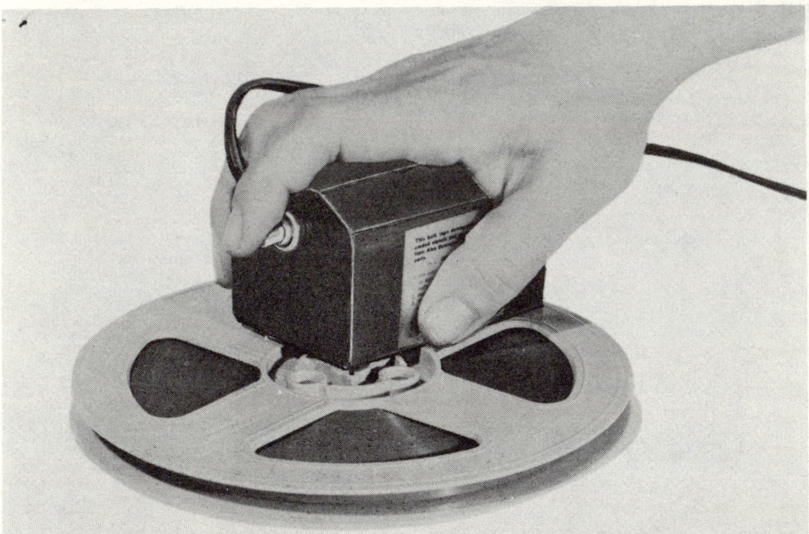

Courtesy Robins Industries Corp.

(B) Handy eraser for recording tape.

Fig. 9-5. Magnetic-tape bulk erasers.

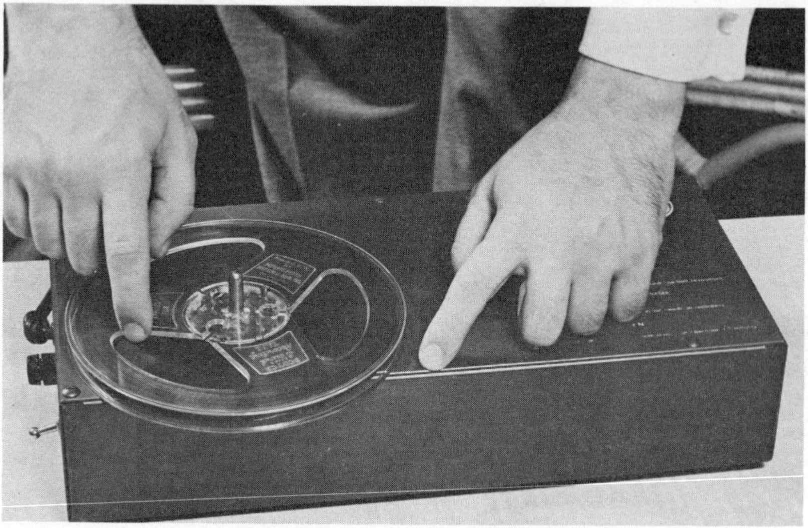

(A) Tape placed on bulk eraser.

(B) Tape being removed from bulk eraser.

Fig. 9-6. Erasing tape with a bulk eraser.

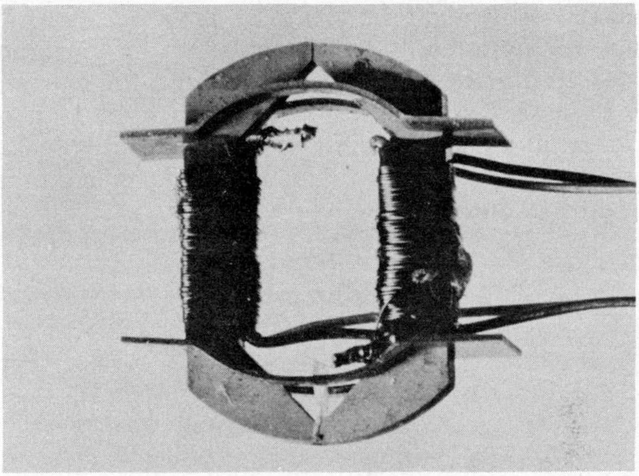

Fig. 9-7. The ring head is a closed magnetic circuit with a gap at one point.

Frequency Response

During the playback operation, the high frequencies on the tape are limited in reproduction by the width of the gap. That is, as the recorded wavelengths on the tape approach the physical size of the head gap, the signal becomes greatly attenuated. In simpler terms, a 7500-Hz wave recorded at 7.5 ips will occupy a space of a thousandth of an inch on the magnetic tape. However, from the discussion of the theory of magnetic recording, we saw that for one wavelength there are two bar magnets laid on the tape facing each other. Thus, each magnet will be only .0005 inch long. Therefore, to properly reproduce a signal of the frequency of 7500 Hz at 7.5 ips, the recording gap must not be any larger than .0005 inch or the signal will be greatly reduced.

Because of the present trend toward extending frequency response at lower speeds, it is not uncommon for several home-type recorders, as well as professional machines, to go up to 15,000 Hz at 3.75 ips. This means that the gap length must not be longer than about .00028 inch long. It is obvious that extreme precision is required in the manufacturing process in order to make a gap of the required narrow width.

One might logically reason: Why not make the gap as short as possible to record the highest frequency possible? The disadvantage of a short gap is that the available signal energy is reduced as the gap length is shortened. In other words, as the gap becomes smaller and smaller, more and more of the magnetic-flux lines flow across the gap instead of around the entire core length which encircles the coil. Of course, the only signal produced is that generated by the flux that passes through the coil (Fig. 9-8).

Therefore, it is desirable to have a wide gap for maximum signal. However, it is desirable to have a narrow gap for maximum high-frequency response. Generally, the gap length is made as narrow as possible to a point where the desired signal output is still obtained. Present gap lengths generally range between 40 millionths and 5 ten-thousandths of an inch long.

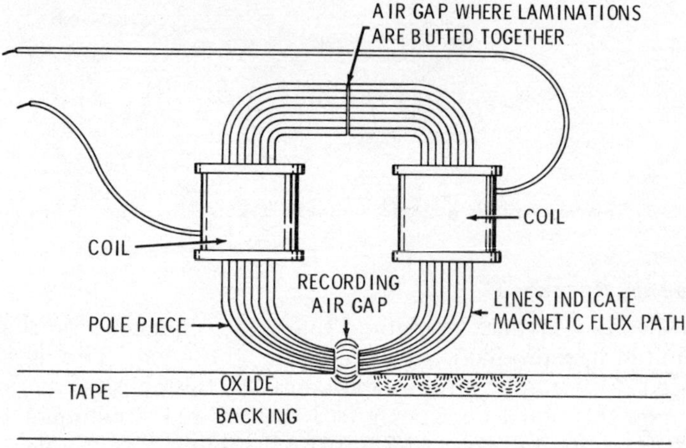

Fig. 9-8. Diagram showing flux through the ring.

The shape of the head at the gap is an important consideration. From the standpoint of wear, the gap should be as deep as possible. Thus, when the tape passes over the head, wearing it down as any sliding friction inevitably will, there should be a good depth of metal before the gap length becomes longer, ruining the high-frequency response. The effects of poor head contact with the tape are shown graphically in Fig. 9-9.

It is obvious that when the head has been worn down as shown in Fig. 9-10, the high-frequency response will be greatly impaired. Therefore, it is desirable to have a deep gap to prolong head life. A deep gap, however, is undesirable from the standpoint of output. A deep gap has the same effect as a narrow gap in that it prevents all the flux from going through the coil, but rather shorts out the flux through the air gap. Again, a compromise is reached in head design between long wear and high output.

A factor that has helped this problem tremendously is the improvement of tape quality, partially the development of smooth, fine-grained tape but scarcely wears down heads at all.

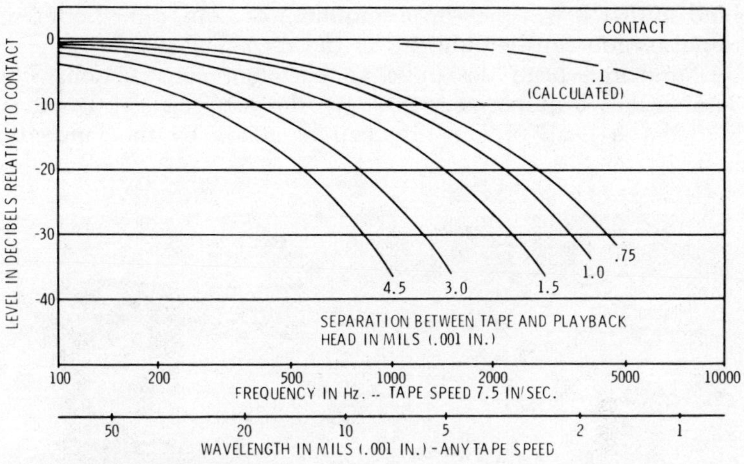

Fig. 9-9. Attenuation caused by poor contact in playback.

Magnetic Head Construction

Three methods of magnetic head construction have developed in turn. One is the lamination method (Fig. 9-11) such as used in most professional recorder heads. In the lamination method, thin laminations are stamped to give the desired gap profile and core structure. The laminations are then stacked together to form a magnetic head.

If two windings are used on the head, and if the head is made small with symmetrical core structure, the hum field can be nearly balanced out. This is especially important in home-type recorders because close proximity of motors and power transformers produces strong magnetic fields which are picked up by the reproduce head. Thus, if a reproduce head can be correctly designed and adequately

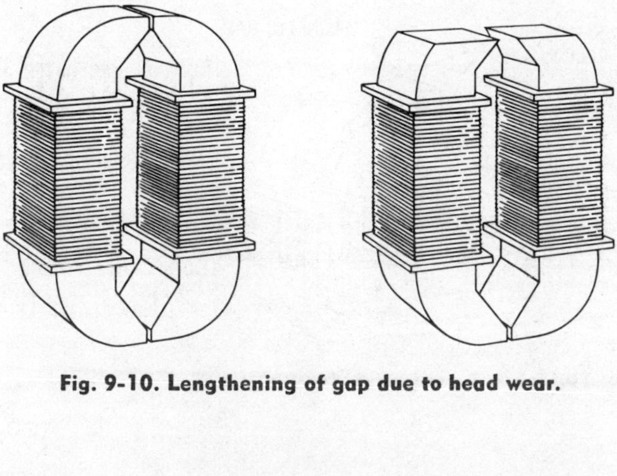

Fig. 9-10. Lengthening of gap due to head wear.

shielded so that it will pick up a minimum of hum, a far better signal-to-noise ratio can be obtained in the recording process.

The laminated head design is an excellent construction. Since the laminations are stamped out, any desired gap depth can be made. Thus, a head of long life can be made by the lamination method.

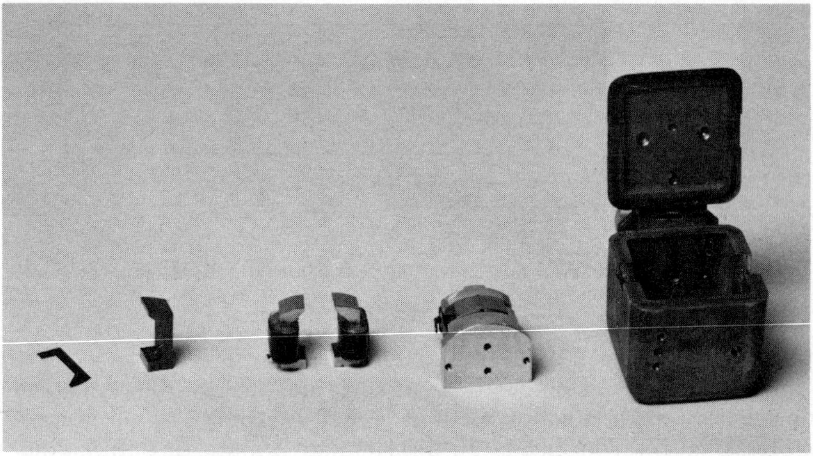

Fig. 9-11. The lamination-stock type head.

The second common type of head construction has a single lamination butted against the end of another lamination to form a gap (Fig. 9-12). The laminations are ground down on the inner surface to provide the necessary gap profile. Then the laminations are in turn inserted between two U-shaped laminations, thus completing the pole structure. This type of construction provides a satisfactory head design.

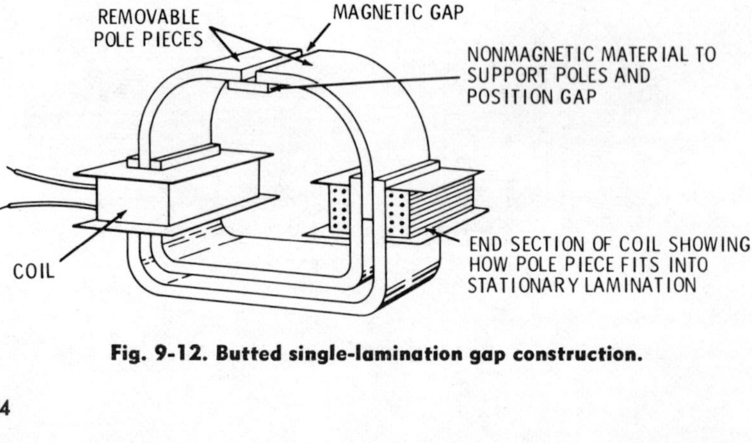

Fig. 9-12. Butted single-lamination gap construction.

It suffers, however, from two disadvantages. The magnetic flux path is rather narrow, thus restricting an easy travel through the core structure, resulting in generally lower output. Also, the point at which the gap occurs is thin, giving the head a relatively short life. However, one big advantage is that the pole pieces are removed and a new gap can be inserted without necessitating the purchase of an entire new head and coil assembly. Thus, when the gap is worn and the high-frequency response is impaired, a new but relatively inexpensive pole piece can be inserted. This pole-piece replacement is generally a factory operation.

The Head Face

The various characteristics of the face of the head have a profound effect on the performance; for example, finishing quality affects the face polish and the cleanness of the gap, and the type of window through which the poles are exposed determines the degree of hum shielding and the low-frequency response curve of the head. The contour or shape of the face influences how well the tape contacts the gap, and the type of face material can aggravate the buildup of oxide deposits on the head face.

The shape of the head face has a definite influence on the methods used for handling the tape and on the resultant performance of the head. The most common face configuration has been the *cylindrical* contour which has a constant radius of curvature of ½ to 1 inch. A felt pressure pad is generally used to insure adequate tape pressure against the gap, or in the absence of a pressure pad a rather high degree of tape tension may be employed from friction brakes on the supply-reel spindle or on the tape itself. The main disadvantage to the cylindrical contour is the shallow radius of face curvature in the gap region.

Hyperbolic face contour is a development in which a sharp (¼ inch or less) radius of curvature is applied to the center of the head face so that the gap line runs down the center of the apex of the curve. Fig. 9-13 shows the comparison between the cylindrical and hyperbolic types of heads. Note that on either side of the sharp nose of the hyperbolic head the face of the head flattens and drops back at an angle of 15° while the normal tape wrap drops back at approximately a 10° angle.

The key to the difference between the two head styles can be seen in the marker lines which indicate the length of tape actually contacting the faces of the heads, about ⅓ as much for the hyperbolic as for the cylindrical shape. The same tape tension will give proportionately more force against the gap of the hyperbolic head. Likewise, when "straight-line" tape travel is incorporated in the tape deck requiring the use of pads, correspondingly less pad pressure needs to be used, which reduces the tendency for tape to "walk" up or down the face of the head.

Current advances in the techniques of manufacture of tape heads have enabled the complex hyperbolic contour to be incorporated into tape heads of all types instead of only the expensive instrumentation and professional classes.

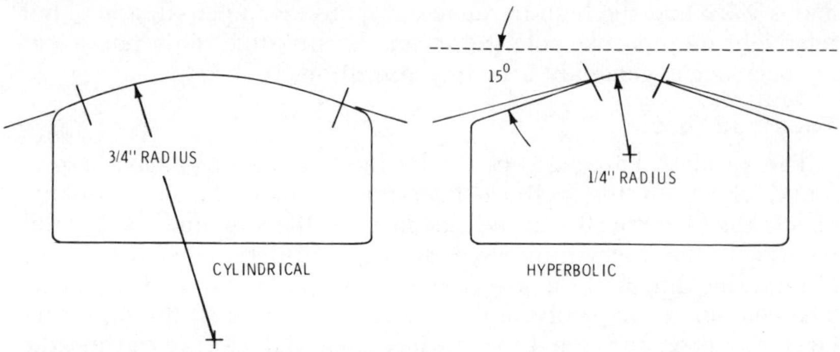

Fig. 9-13. Cylindrical and hyperbolic types of heads.

Face Material

Plastic-faced heads have been used for many years and can give excellent results provided care is exercised to maintain the face of the head free from tape oxide deposits by frequent cleaning. The newer all-metal faced heads have proved to be incomparably superior in their freedom from oxide buildup and are especially desirable for use in background music and other installations where repeated head cleaning is inconvenient. Static electricity due to tape friction, with resultant sticking of the tape, is also eliminated by the metal face. The combination of the all-metal face and hyperbolic contour giyes excellent performance characteristics.

Caution should be exercised in the use of the all-metal heads for high-speed tape duplicators which have bias frequencies of 300 kHz and up. Eddy currents induced in the metal structure by these extremely high bias frequencies can cause low-efficiency and heating effects. For such applications duplicating heads of plastic construction are advised.

Frequency Response and Output

In magnetic head construction the pole pieces should be long enough to reproduce low frequencies, and the gap should be sufficiently short to reproduce high frequencies. In brief, this is the main criterion for frequency response in the magnetic head.

As mentioned earlier, to obtain a satisfactory output from the tape, the magnetic-flux path should be large through the coil area and should be very small at the gap area. Thus, it is desirable that the flux travel through the coil very easily, yet jump over the gap with difficulty. Consequently, the major portion of the flux will pass through the coil rather than the gap. This provides for high output.

One might logically ask: "Why cannot a great many turns be wound on a magnetic head to provide high output?" Of course, the more turns that are wound on a magnetic head, the greater will be the output voltage. However, when more than a certain number of turns are wound on the head, the capacity in the winding becomes so great that it restricts the high-frequency response.

In many head designs, the resonant point is at the top end of the audio-frequency range, now occurring near 15,000 Hz. Thus, if any more turns are added to the head, it is impossible to reproduce the high frequencies since the resonant point is lowered. Therefore, to obtain good high-frequency response, no more than a certain number of turns can be put on the head.

Some heads serve as a low impedance device. Using a head with relatively few turns, the signal can be fed to a transformer which steps it up to a high impedance output. Fig. 9-14 shows a circuit using this method. However, the capacitance of the transformer combined with the capacitance in the head yields nearly identical results as using a high-impedance head. The high-frequency response, then, also limits the output of the head.

Several methods have been proposed to overcome this deficiency. One of the most novel methods is to design a head that feeds a transformer with two different turns ratios. That is, a head feeds a filter network which separates the low and high frequencies. The low frequencies are fed to the transformer, which has a great turns ratio, providing high output at low frequencies. The high frequencies, however, are fed to a lower turns ratio which does not suffer from capacitance. From the chapter on equalization, it can be seen that this technique gives the desired effect since the low frequencies are reproduced at far less amplitude than high frequencies. By properly designing this network, it is possible to obtain all the equalization in the input stage and obtain high signal level. However, in most recorders, the conventional head feeds a very high-gain amplifier with as low a noise as possible after which the equalization is applied.

The advent of transistorized amplifiers has changed this considerably. Transistor input circuits can match any impedance value by making appropriate modifications. Thus, the designer can design the head to suit the recording amplifier output, then design the playback amplifier input to suit the head.

Magnetic-Head Wear

Magnetic-head wear has long been considered a great problem by many people. It is true that in early recorder designs, head wear was a major problem. However, with the introduction of lubricated magnetic tape in 1949, the problem of head wear was greatly reduced. Modern magnetic-tape construction incorporates a lubricant which will generally last the life of the tape.

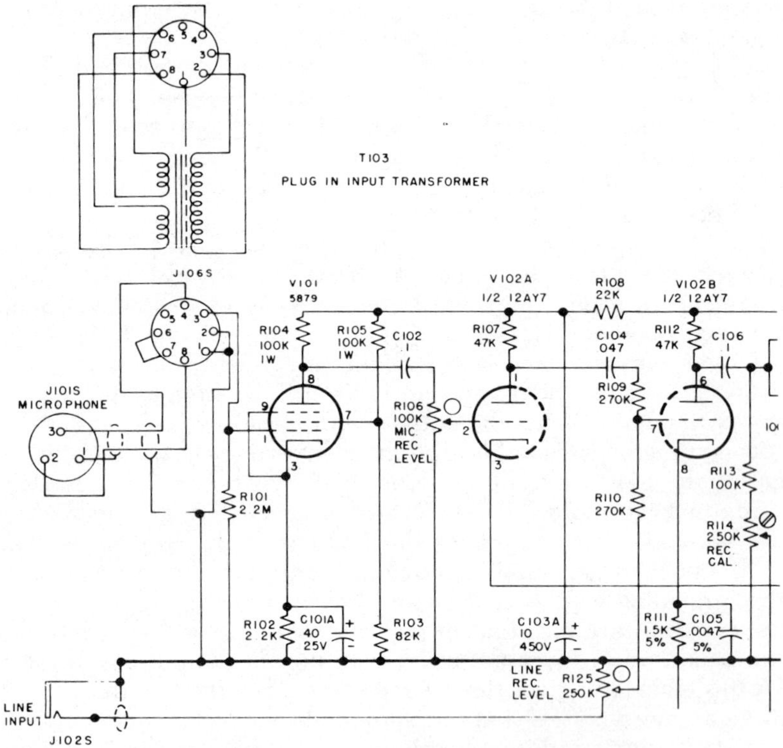

Courtesy Midwestern Instruments Inc.

Fig. 9-14. The Magnecord low-impedance input circuit.

Many misconceptions surround the problem of head wear. Most people feel that head wear is due solely to the fact that the iron-oxide tape coating is extremely hard while mu metal is soft. It is true that the iron-oxide coating of magnetic tape is almost twice as hard as the mu metal used in the recording head. The slipping friction of the coated iron-oxide particles against the soft mu metal head may seem at first glance to be abrasive. However, the contact of the hard tape against the soft head is analogous to the action of a bearing.

In bearing construction, the shaft is generally hard and the bearing material itself is soft. This means that the shaft rotating within the bearing runs smooth and free. The lubricating film between the shaft and the bearing provides good protection, assuring an almost indefinite life as long as the lubrication is there. However, it is the dirt that works into the bearing that causes wear, necessitating replacement. The same is true for magnetic tape. It is the dirt that collects on the tape which tends to scour and abrade the head far more than the iron-oxide coating.

The magnetic head, being considerably softer than iron oxide, serves the same function as the soft bearing. The hard iron-oxide tape coating can be compared to a steel shaft turning on the bearing. Thus, it can be seen that a mu metal head sliding against magnetic tape provides a reasonably good combination for long head life.

Many materials harder than soft mu metal have been explored as a possible head replacement. Ferrite core heads are becoming more popular because ferrite is a very hard material, more than twice as hard as mu metal. Precision manufacture of the head, coupled with the super-smooth tapes now available, has solved what was for a long time a serious problem.

As the head wears, several changes occur in the operational characteristics. These changes, which take place on a gradual continuing basis, are:

On Record

1. Peak bias-current requirement decreases.
2. Inductance decreases.
3. Actual bias current increases.
4. Record sensitivity increases.
5. Record current increases slightly.
6. Gap linearity frequently decreases and finally the gap abruptly opens.

On Playback

1. Playback sensitivity generally increases.
2. High-frequency response relative to low-frequency response increases and then abruptly reverses as the gap opens.

In the record process, 1 and 3 aid each other to reduce the high-frequency system response by producing an overbias. In many home-type machines, a bias control is not provided; therefore, good performance cannot be maintained by compensating for the high-frequency losses.

HEAD LIFE

Head life is highly dependent on the design factors of the recorder involved. The differences in head wear with and without pressure pads can sometimes be a factor of 10. If the recorder uses relatively small-area pressure pads, and if the pressure-pad force exceeds one-half ounce, head life can be as low as 300 or 400 hours of operation. On the other hand, if the recorder does not use pressure pads, the tape guides are smooth, and the depth of the metal at the gap is sufficiently large, 3000 to 4000 hours of operation can be expected.

PRESSURE PADS

Intimate head contact is essential in attaining high-frequency response. The professional recorders use a high tape back tension that figuratively stretches the tape across the head, holding it in contact by tension.

However, in home machines, where very little if any back tension is applied, pressure pads are used to hold the tape against the head. The pressure pad is generally a good device to assure intimate tape contact with the head. However, the pressure pad tends to wear the head somewhat unevenly, developing pits and craters in the head as shown in Fig. 9-15. This will eventually lead to poor tape conformity to the head and loss of high frequencies. Of course, pressure pads themselves are subject to wear and require occasional replacement as shown in Fig. 9-16.

TAPE GUIDES

When tape passes across the head, it is essential that it is accurately guided, enabling it to pass over the head in a straight rather than a zigzag manner. The tape must be guided to close tolerances so it will attain the same path on each playback. For example, if the tape is poorly guided, it tends to weave back and forth across the head. The head is then thrown out of alignment with respect to the tape, reducing the high frequencies or causing severe amplitude variations.

A number of guiding systems are employed. Most of them have restricted guides on each edge of the tape placed before the record and reproduce head to guide the tape in a straight path, as shown in Figs. 9-17 and 9-18. Several professional recorders employ glass tape guides.

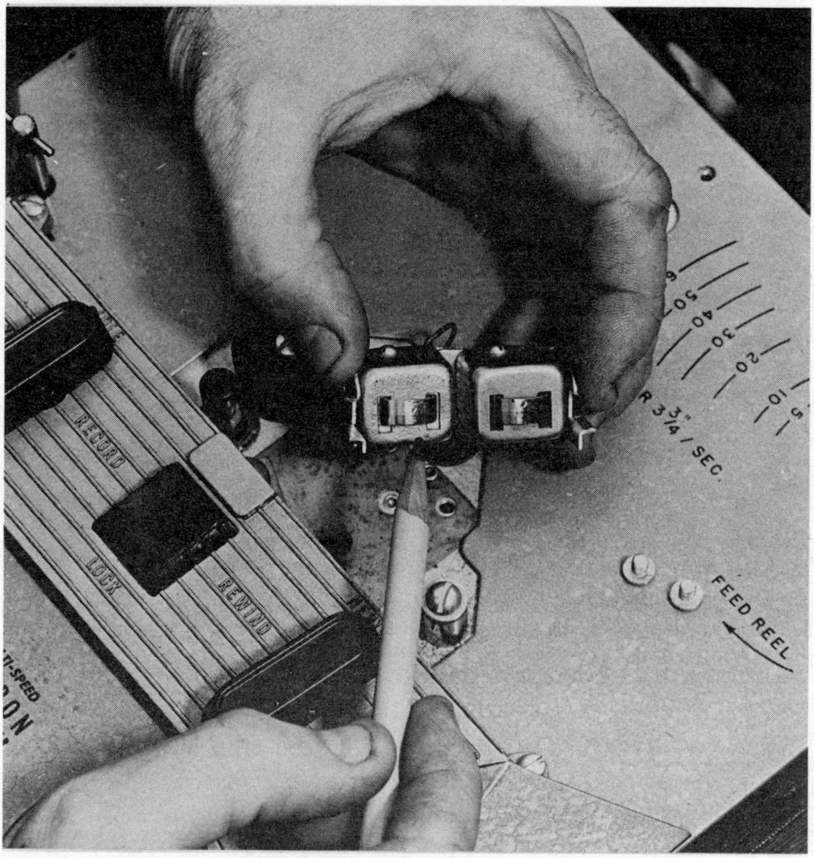

Fig. 9-15. Pitting caused by worn pressure pads.

HEAD ALIGNMENT

As we have seen, when the tape is poorly guided it is thrown out of correct azimuth alignment. In correct azimuth alignment, the gap in the head should be exactly perpendicular to the tape, assuring that tapes recorded on one machine will reproduce properly when played on another.

When the same head is used for both record and playback, the azimuth can be out of perpendicular alignment, and the tape can still be reproduced satisfactorily since both heads will have the same angle. However, for interchangeability of tape, it is vital that the azimuth be correctly set.

Excellent azimuth alignment tapes can be obtained from several companies. These tapes have a high-frequency signal at short wave-

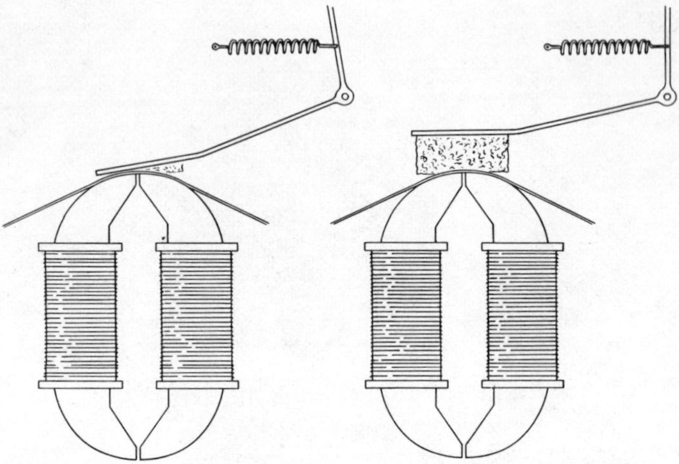

Fig. 9-16. New pressure pad compared to worn one.

length recorded on the tape at a very precise 90-degree angle from the edge. The alignment tapes are accurate within several minutes of arc. When adjusting an alignment on a machine, the normal procedure is to play the tape, then adjust the reproduce head for maximum output as shown in Fig. 9-19. Care should be taken to reach

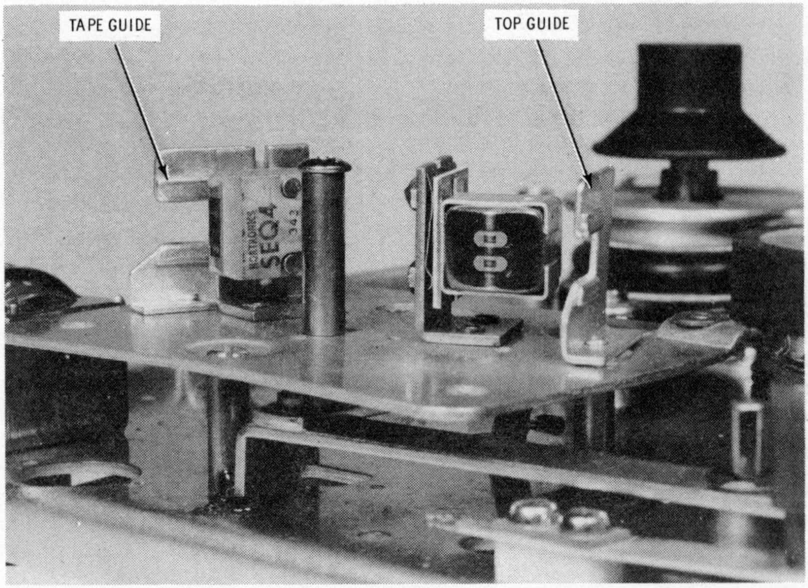

Fig. 9-17. Tape guides used in Ampex Model UST-4 recorder.

Courtesy Ampex Audio Inc.

Fig. 9-18. Recessed rollar pre-positioning tape guide.

Courtesy Ampex Audio Inc.

Fig. 9-19. Adjustment of playback head using a standard alignment tape.

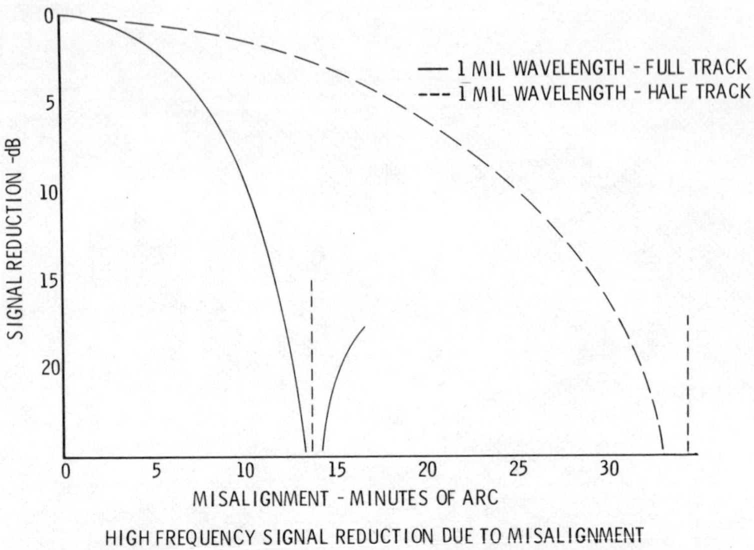

Fig. 9-20. Graph showing importance of head alignment.

Courtesy Ampex Audio Inc.

Fig. 9-21. Adjustment of record head.

the point of maximum output since a lesser peak in output will occur on each side of the maximum-output position. The illustration in Fig. 9-20 shows the effect of head misalignment on the output.

When maximum output is obtained, the record head, if a separate record head is used, can be adjusted as shown in Fig. 9-21 by recording a high-frequency tone on the tape of approximately one mil wavelength, where:

$$\text{wavelength} = \frac{\text{tape speed}}{\text{frequency}}$$

Then the azimuth of the record head can be adjusted until the maximum output from the reproduce head is obtained. Since the reproduce head has been previously set to an accurate azimuth, the record head will also be set accurately.

NUMBER OF TRACKS

The original terms "full-" and "half-track" recording were sometimes referred to as single and dual tracks. Dual track meant, of course, that two tracks were not recorded, but rather half of the tape width was used to record a single track instead of the full width of tape. A comparison of the heads used for full-, half-, and quarter-track recording is given in Fig. 9-22.

Fig. 9-23 shows a head assembly, with associated electronics, used on a professional multitrack machine. Much discussion was devoted to the merits of full- versus half- and quarter-track recording. Considering the attention and controversy which have surrounded this subject, it is strange that misunderstandings still remain. A persistent folklore abounds which is unsubstantiated by facts.

In theory, a full-track recording should give twice the output voltage of a half-track recording. Or, it should give a 6-dB better

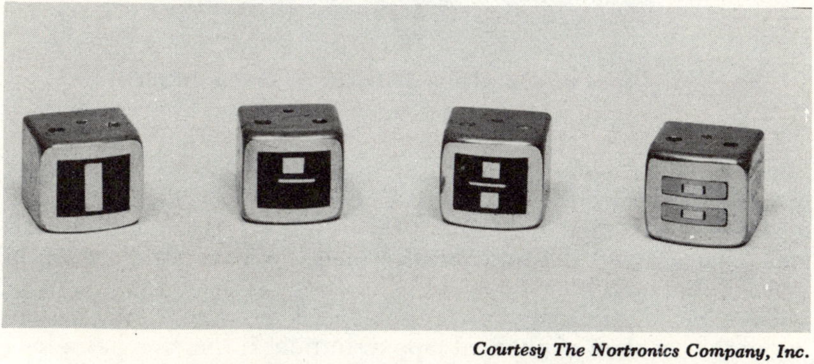

Courtesy The Nortronics Company, Inc.

Fig. 9-22. Various types of modern record-playback heads.

signal-to-noise ratio, assuming the noise occurs in the amplifier, as it generally does.

However, when making a narrower width gap head required for half track, generally a better and more efficient head is obtained. Therefore, in practice, the difference in output between a half- and full-track recording is usually only about 3 dB. Consequently, one may expect approximately a 3-dB better signal-to-noise ratio using a full-track recording rather than a half track. Correspondingly, as tracks are multiplied in number, the signal-to-noise ratio deteriorates.

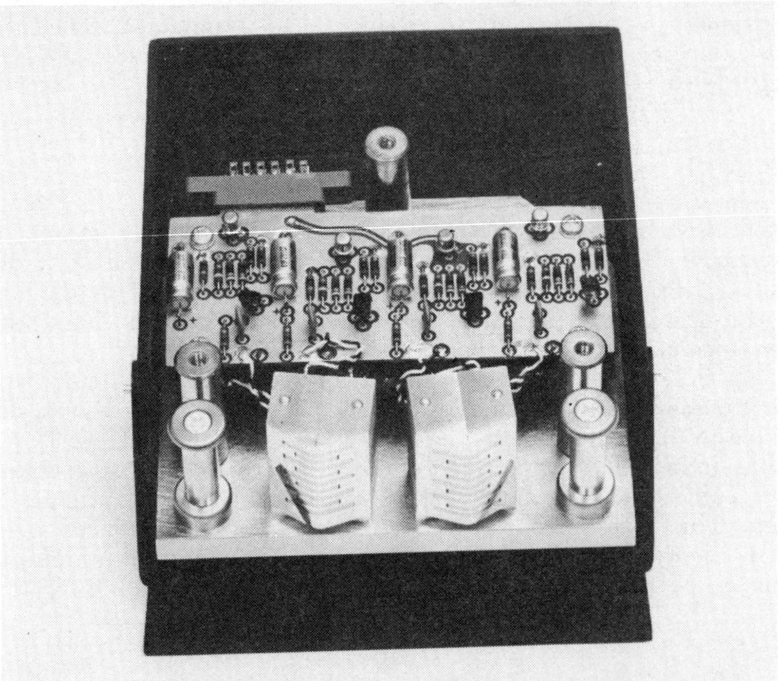

Fig. 9-23. A head assembly for professional use with multitrack tapes.

The economy in tape is, of course, obvious with a half- or quarter-track recording, obtaining twice the recording time on a single reel of tape. Also, since the gap width is shorter on half-track and quarter-track heads, they are less critical in respect to azimuth adjustment than the old full-track head, which is still another feature in favor of more tracks.

The chief disadvantage of the multitrack head recording system lies in the problem of physical tape distortion. If the tape has a long edge or has a wavy surface, use of the multitrack head will affect performance more adversely than a full-track head. When the tape is physically distorted, the long edge of the tape will tend to lift away from the head. In recorders using a full-track head, this is often not likely to be serious. In the case of multitrack recording, if the tape should lift up on one edge, and if this is the edge upon which the recording is being made, severe amplitude fluctuations will result.

Contrary to popular belief, the frequency response is not directly affected by the track width. Only the output or signal-to-noise ratio is affected. As was previously explained, in theory, a 6-dB better output is obtained by using a full-track head as opposed to a half-track head. However, due to the more efficient construction made possible with half-track heads, a full-track head can generally reproduce only 3-dB more signal than a half-track head. The frequency response in either case should be identical, assuming identical gap lengths and construction of the heads.

Since 1956 stereophonic recording has been available. The first commercial stereo recorder used two staggered heads with each head using one track on a tape traveling at 7.5 ips.

About a year later, a single head with two magnetic elements was introduced to simplify the tracking and alignment problems found in the use of two separate heads. The heads were rapidly improved to the point where it is now possible to record two complete stereophonic programs on quarter-inch tape. The four-track stereo heads on the market at present make it possible to record with adequate fidelity even at the 1⅞-inch speed.

HEAD REPLACEMENT

Each recorder head requires a different amount of bias for maximum performance. Generally, at the time of manufacture of the recorder, the bias is either adjusted or permanently set to fall within a region of proper operation when used with a given head. In professional recorders with adjustable bias, when replacing a head on a machine, the bias should be always readjusted to give optimum performance.

As was noted in the chapter on bias oscillators, the bias must be adjusted properly, for if it is too low, serious distortion will result. If it is too high, the high frequencies will be lost. Therefore, a head of another manufacturer generally should not be put in a machine designed for a specific type of head since the bias requirements are likely to be different, unless, of course, care is taken to properly adjust the bias for the new head.

Heads also vary as to the amount of equalization necessary, especially at high frequencies. Therefore, if changing heads to those made by another manufacturer, not only the bias, but also the equalization often will need readjustment. However, on most home-type machines, the head can be replaced with the head of the same manufacturer without difficulty. The only precaution that must be observed is to be sure that the head is properly aligned in respect to azimuth.

In addition, replacement heads are available from other sources. Fig. 9-24 shows a head assembly designed for replacement purposes. The old heads are removed and the new assembly, which must be designed for the specific recorder, is inserted. The use of an assembly such as this permits the updating of old recorders with newer-type heads.

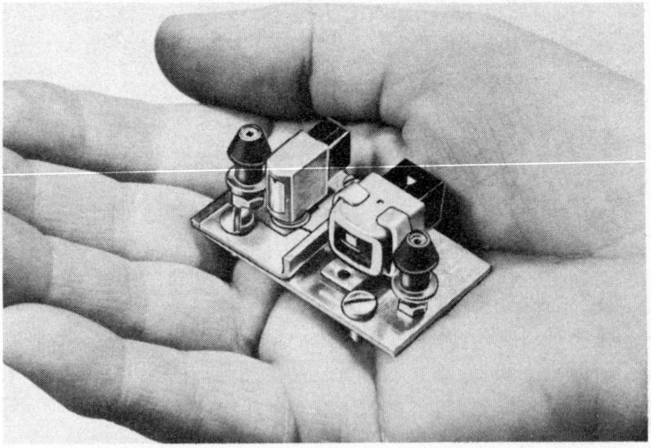

Courtesy The Nortronics Company, Inc.

Fig. 9-24. A replacement head assembly.

CROSS-FIELD HEADS

One way of applying the cross-field technique was described in Chapter 7. This procedure uses heads on opposite sides of the tape, with the bias head behind the tape and the record head on the coated side. With this method, precise placement of the heads is very important so that their gaps will align on opposite sides of the tape.

A better solution would be to lock the two heads together in a single assembly so they cannot shift, but this is impractical when they are located on opposite sides of the tape. This problem prompted the invention of the single side cross-field head.

This is a single head that contains separate gaps for the bias and audio frequencies; these gaps are close together lengthwise along the tape. This method works by changing the field contours in a depth dimension relative to the tape surface. Fig. 9-25 shows the arrangement and also how it changes the field penetration.

This modification, by enabling better penetration of the tape, particularly at the highest recorded frequencies, overcomes the previous conflict between bias adjustment for minimum distortion and maximum high-frequency response.

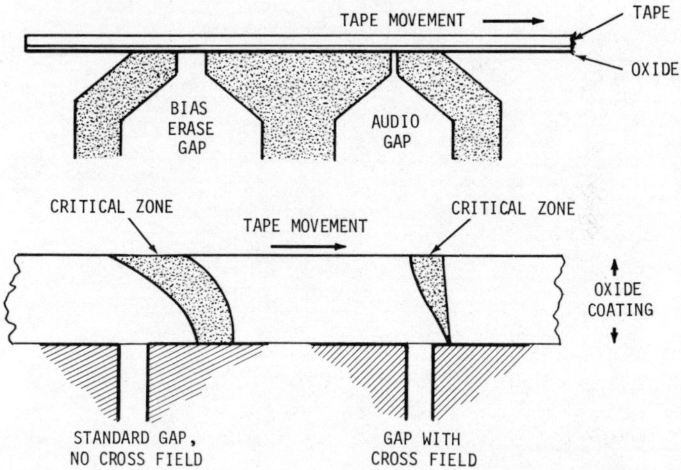

Fig. 9-25. Cross-field magnetization using a single head assembly.

CLEANING HEADS

It is important to keep magnetic heads free from any deposits of dirt oxide, or adhesive which may have oozed from poor splices. A lens paper, a soft cloth, or a cotton applicator is satisfactory for head cleaning.

The use of solvents for cleaning purposes should be restricted to products specifically manufactured for head cleaning. These products will not react in any way with the resins, plastic surfaces, or metal surfaces of the head. This same cleaner should be used on the tape guides, capstan shaft, and pressure rollers. A second solution is a silicone that dispenses a thin coating of lubricant on the cleaned parts.

Special tape care kits are manufactured and sold by various manufacturers, such as Nortronics Company and Robins Industries, and are excellent for cleaning and lubricating purposes.

CHAPTER 10

Magnetic-Recording Tape

Present-day magnetic-recording tape is very different from the early pre-World War II German magnetic tapes. As we saw in an early chapter a more potent oxide, a form of gamma Fe_2O_3, was developed. The new oxide displayed a 14-dB superiority in output and vastly improved frequency response over German tapes. Known as *red oxide,* the new type of magnetic oxide was quickly accepted as the international standard for magnetic tapes and films.

Thanks to the superior magnetic properties of red oxide, coupled with basic improvements in head design, recording speeds were reduced from the initial 30 ips to 7.5 and 3.75 ips with excellent results. Further inprovements resulted from refinements in processing and from introduction of chromium dioxide.

TAPE MANUFACTURE

The complete magnetic tape is actually composed of many substances, each making an important contribution toward improved physical and magnetic performance. The basic recording material is a magnetic oxide. Several forms of magnetic oxide will be discussed.

Magnetic tape is generally made by coating the oxide dispersion, combined with a binder material, on a wide web of plastic backing. Again, a wide variety of backing materials are in use at the present time, several of which will be examined later in this chapter.

The oxide is made to adhere to the base material by means of an adhesive or binder incorporated in the tape construction. The binder serves two purposes. One, it holds the iron-oxide particles tightly together in as dense a mass as possible. Two, it holds the oxide firmly to the base material.

The binder also carries a plasticizer, used to keep the tape soft and supple, and a lubricant to help the tape pass the guides and magnetic head with a minimum of friction and wear.

The wide web of base material coated with magnetic oxide is then slit into a variety of widths and lengths. The most common sizes for home and professional recording studio application are the quarter-inch width and the .015-inch width used in cassettes. However, telemetering, video tape recording, and other special applications require tapes that are slit to a variety of widths up to a foot or more wide. Magnetic film is also available in the standard motion-picture sizes which are 35, 17.5, and 16 mm. The base is identical to motion-picture film. However, instead of being coated with a photographic emulsion, a magentic iron-oxide dispersion is used. Also, picture film is "striped" along the edges, after the picture is pressed, to receive the magnetic sound tracks that many of the new wide screen types of movies use.

The quarter-inch tape has always been available in a variety of standard lengths, with reel sizes of 3, 4, 5, 7, 10.5, and 14 inches in diameter. The newer "extra play" magnetic tape offered 50% additional recording time, and later developments extended time some more by using even thinner tape. Some of the various sizes of magnetic tape available are illustrated in Fig. 10-1. The manufacturers

Courtesy Eastman Kodak Co.

Fig. 10-1. Some of the various available sizes of type.

generally supply more than the labeled footage to allow for threading and to serve as a "cushion" on the reel hub. Thus, a 1200-foot reel generally has about 1240 feet.

Backing Materials

The early German tapes used paper as a base material. Paper has several desirable characteristics in that it is quite stable in respect to temperature and humidity changes. However, a big physical limitation to paper-base magnetic tape is that it is fragile and not too pliable or flexible. It will wear easily. Paper tape has also a rough and nonuniform surface. In magnetic-tape coating, the uniformity of coating is extremely important to obtain a consistent output from the tape. If a magnetic oxide dispersion is coated on a rough surface, it naturally will adhere to the contour of the peaks and valleys of the backing surface.

To obtain a more uniform coating surface with greater potential frequency response, a plastic-base tape was introduced. The first base used was cellulose. It possessed several excellent characteristics, having the smoothest surface of any plastic tape then developed. Consequently, it gave the smoothest magnetic coating which, in turn, yielded a more uniform output.

Cellulose acetate is, also, surprisingly strong. The standard quarter-inch tape has a tensile strength of 5.5 pounds. The new extra-play magnetic tape, using a 30% thinner acetate backing, has a tensile strength of 3.4 pounds. Considering that the maximum tension exerted by tape recorders with proper use does not exceed a few ounces (including the sudden stress on the tape when braking and reversing), cellulose acetate is relatively rugged and holds up well.

The main disadvantage of cellulose acetate is that it is not stable with temperature and humidity. Thus, when the humidity goes up, the tape tends to expand. When the humidity falls, the tape has a tendency to shrink. Expansion and contraction physically distort the tape.

The ideal backing should have a high strength without stretch, a very low coefficient of humidity expansion, be stable over a wide temperature range, have a perfectly smooth surface, and cost very little. Unfortunately, however, the early backings did not combine all these ideal qualities in one tape backing. Then came a newer base material, showing great potentialities, known as polyester film. Its common name is *Mylar,* a registered trade mark of E.I. du Pont de Nemours & Co. Inc.

The polyester backing has some very interesting properties, making it especially suited for tape recording. It is very stable with respect to temperature and humidity. That is, the temperature and humidity can vary widely and it will retain its original shape and size. It has greater strength than acetate with little yield stretch. And it still has a relatively smooth surface.

The outstanding property of acetate, accounting for its wide use for a long while, is its excellent surface and thickness uniformity. Early polyesters suffered in surface uniformity when compared to acetate. However, recent production advances yielded a surface smoothness of polyester equal to that of cellulose acetate.

Magnetic-tape backing exhibits both a yield point and a break point. Like bending a green twig, the yield point is the point at which the tape upon being stretched will not return to its original length but remains partially elongated and misshapen. When a tape is stretched beyond its yield point, the recording on it will be distorted and the tape will be damaged. The graph in Fig. 10-2 illustrates the yield point of various backing materials. When tension reaches the break point, the tape will snap in two. It is better for the magnetic tape to snap than to approach its yield point without breaking, thus distorting the recording.

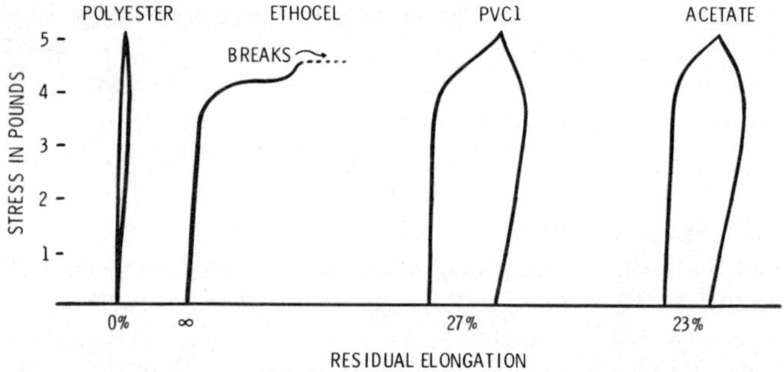

Fig. 10-2. Comparison of backings used on magnetic tapes.

Polyester at 150 gauge exhibits a very excellent stretch recovery, showing no dead stretch up to five pounds tension per quarter inch. By contrast, 140-gauge acetate shows a 23 percent residual elongation at five pounds stress. However, as we have already seen, any consideration of backing strength is largely academic because tensile strength is more than adequate for the tape handling mechanisms that exist today. The additional strength of polyester, while giving an extra margin of safety, is seldom required.

New Tape Construction Trends

An interesting trend in tape construction is toward the use of thinner and thinner backing material to obtain greater recording time. Reducing the thickness of the backing material, coupled with a reduction in the thickness of the oxide coating, does not, surprisingly enough, greatly affect its strength. For example, reducing the backing thickness of cellulose acetate from 140 to 100 gauge, or by 30%, reduces its strength by only 15%. Consequently, tapes of thinner backing material can readily work on all recorders with no particular problem.

The thinner backing possesses several advantages in that it conforms more readily to the head, resulting in greater frequency response. Any magnetic-tape backing should be limp and flexible, enabling the tape to conform perfectly to the magnetic head contour. A thick backing material tends to contact the head improperly, causing amplitude variations. Therefore, a thinner backing material gives better high-frequency response with more uniform output.

In addition to cellulose acetate and polyester, there are several other plastic films. They are ethyl cellulose and tensilized polyvinyl chloride, more commonly known as *Luvitherm. Luvitherm* is used extensively in the German magnetic recording industry. Like polyester, the outstanding characteristic of *Luvitherm* is its low coefficient of humidity expansion. *Luvitherm,* however, is limited by upper operating temperature, undergoing deterioration at just about 150 degrees Fahrenheit.

The Eastman Kodak Co. has developed a special acetate known by the trade name of *Durol. Durol* combines the strength of *Mylar* and the low-strength advantages of other acetates in one tape.

Magnetic Oxides

The early German tapes used a type of oxide known as raven red. This term was conveniently borrowed from the barn paint of the same name. Literally a high grade of barn paint which was painted on a paper backing, the raven-red oxide had lower output, especially at high frequencies. A substantially higher tape speed was required to attain a satisfactory signal-to-noise ratio and frequency response.

However, with the introduction of black-oxide tape in America, a tremendous increase in output was attained with greatly improved frequency response. The black oxide proved to be an excellent recording medium. However, it possessed the undesirable property of being extremely difficult to erase because of its memory effect. That is, once the tape was erased during subsequent recordings, the supposedly erased recording would tend to reappear in the background and ruin the new recording.

In order to develop a tape that would avoid this difficulty, and yet maintain good high-frequency response and high output, the present "red oxide" was developed. It became a standard in the recording industry.

A more recent development has been the introduction of high-output oxide. This oxide possesses even greater output (8 dB) than the conventional red oxide for a given distortion, as shown in the graph in Fig. 10-3. It is especially useful where noise is a problem in the reproduce process. For example, in equipment where the close proximity of the power transformer and drive motor to the reproduce head induces serious hum, the use of high-output tape with 8-dB more output will require 8-dB less amplifier gain, reducing noise by that amount.

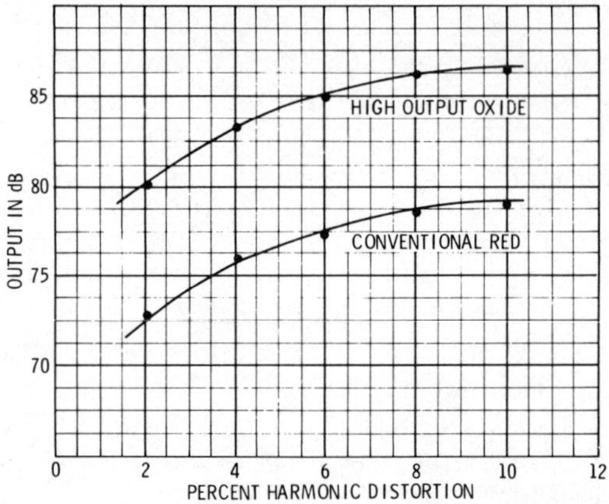

Fig. 10-3. Performance curve of high-output tape compared to conventional tape.

Also, in self-powered portable recorder applications, the extra sensitivity of the high-output oxide (4 dB) means that the tape is especially sensitive to weak sounds. It is also much less likely to overload and distort when subjected to unexpected and sudden loud sounds. Thus, high-output tape provides an extra margin of safety, freeing the operator from the necessity of constantly riding the gain control.

Even more recent has been the development of chromium dioxide, which is another black tape. This can record at another 8-dB higher level and has a much higher coercive force, so it is much harder to erase, like the earlier black tape. However, with proper attention to levels of audio, bias, and erase, it far outperforms earlier tapes, with better frequency response and almost no print-through or deterioration with storage.

Binder Formulations

The binder is an adhesive that is blended into the magnetic dispersion, making the iron-oxide particles adhere to the backing. Also incorporated in the binder are a number of other elements such as lubricants, preservatives, and plasticizers.

The binder is designed so that it will not be too hard or brittle. If it is too brittle, when the tape rounds a sharp corner, the coating will actually crack and the oxide will chip off. The binder should also age well, withstanding changes in temperature and humidity, retaining its bond to the backing. It is equally important that the oxide not be so soft that it will easily rub off during recording or playback.

Certain types of binders become tacky at high temperature, causing the tape to stick and squeal as it is propelled past the record head. The result is serious high-frequency flutter.

Other tape binders tend to become stiff and hard, curling with temperature changes. Still other binders have a different expansion rate than the backing material, again causing the tape to cup and curl with a change in temperature or humidity. Therefore, it can be seen that the binder as well as the tape backing plays a critical role as to whether or not a tape will cup and curl. The tape must, of course, lie flat against the head, conforming readily at all times, for satisfactory frequency response and uniformity of amplitude.

Tape Lubrication

The tape should be lubricated to improve head wear and reduce friction vibrations in the tape, causing high-frequency flutter. Lubrication will prolong the life of both the tape and the record heads and guides, as well as giving better recording results.

The lubrication is incorporated in the tape binder itself so that it will last the life of the tape, not wearing off as it would if only the tape surface were lubricated. Included in the binder are a number of other properties such as plasticizers that are employed to keep the tape soft and flexible throughout its life. While binder chemical formulations are closely held trade secrets, also included in the binder are agents which permit tape operation at high and low temperatures and improve the physical performance of the tape. One tape manufacturer uses a silicone lubricating process which actually coats the record head with a thin film of protective lubricant.

The processing of modern tapes also includes "calendering," a surface polishing that reduces tape/head friction almost to the vanishing point. It also minimizes friction around tape guides.

LIFE OF MAGNETIC TAPE

Earlier there was some dispute about the permanence of tape recordings. It had been observed that extremely high frequencies seemed to deteriorate with storage. However, parallel observations of natural materials suggested that such deterioration was impossible once the program was impressed on the tape, unless it was deliberately removed.

What those observations ignored was the minute size of the artificial magnetization at extremely high frequencies, not duplicated in nature. While the argument persisted, some believed that the loss was imagined because better quality tapes had become available during the interval over which the tapes were kept. Finally this argument was laid to rest by using phonograph records as the storage medium for the reference.

A high quality phonograph record was transferred to tape and the two recordings compared. At a later date, the same equipment was used to transfer the same program to another piece of the same tape, and the two tapes were compared; it was not imagination—the high frequencies had deteriorated with storage.

This is a feature in which the new chromium dioxide tapes offer much superior performance; the high frequencies do not deteriorate with keeping.

Storage Precautions

While the magnetic properties of recording tape are very stable over long periods of time, care must be exercised to avoid accidental erasure by exposure to strong magnetic fields. Permanent magnets and strong electromagnets will very likely cause erasure or partial erasure if placed within a few inches of the tape. This is the principle used in the so-called bulk erasure process in which an entire reel of tape is demagnetized without unwinding. Since the field necessary to produce erasure must be so intense, it is unlikely that accidental erasure would occur from close proximity to ordinary electric house wiring. Another possibility has arisen since the increase in aircraft hijacking has resulted in airports installing security surveillance equipment. Some of this employs fairly high intensity alternating (usually high frequency) magnetic fields, which can wipe tapes passed through the sensors. Packages should either be clearly labeled or packed in magnetic shielding cases and labeled.

All manufacturers state that, ideally, magnetic tape should be stored in a room with relative humidity being maintained between 40 and 60 percent with a room temperature of 70° Fahrenheit. However, tape storage under these carefully controlled humidity limits is a ridiculous condition for magnetic-tape users to meet. An air conditioning unit that could control room humidity within such a range would cost many thousands of dollars and very few such units exist in the world today.

Realistically, storage of tape at moderate temperatures such as maintained in the home is desirable. Tape that is subjected to extremes of temperature for short periods of time such as in shipment is generally not affected. The same precautions exercised in storage of safety-base film can be applied to magnetic tape. If the tape has been subjected to extreme temperature, allow it to return to room temperature before using. High temperature plus high humidities, however, tend to rapidly deteriorate the cellulose-acetate backing of magnetic tape.

The worst condition is high temperature with low humidities. The backing will tend to dry out and become brittle. Embrittlement of tape will result from exposure to prolonged low humidity which exists in heated areas during cold winter months. Just as woodwork shrinks in winter and swells in summer, the same principle applies to magnetic tape. In winter, low-temperature air is forced into a room and is heated. It can hold more moisture and consequently draws moisture from all objects in the room. Subsequent storage at normal humidities completely restores the desired tape properties, the tape picking up its original moisture and becoming limp again.

If the relative humidity is subject to large variations, magnetic tape can be safely stored in sealed metal cans. Seal the cans with a pressure-sensitive plastic or cloth-backed tape if long-time storage is anticipated. The use of desiccants or humidifying agents is not recommended because of the difficulty in controlling the results.

The tape should be stored, wound loosely on its reels. Many professional recorders wind the tape far too tightly on the reel during rewind or in the normal forward position. Thus, when the tape goes through several humidity cycles expanding and contracting, the pressure and tension built up severely stretch and deform the tape. To avoid this condition, the tape should be rewound with very low tension before being stored. Before using a tape that has been stored for any appreciable length of time, the tape should be rewound through a recorder to relieve any tensions that may have set in while the tape was wound on the reel.

PRINT-THROUGH PROBLEMS

When tape is wound on a reel, the magnetic field from one layer tends to somewhat magnetize the adjacent layers of tape, causing a signal transfer. This is called print-through or crosstalk. The signal transfer can be heard either ahead or behind the recorder signal, depending on which layer the transfer came from. Often both the preprint and the postprint signals can be heard.

Print occurs mostly in staccato-type recordings in the middle frequencies, such as when suddenly a voice is heard, then stops, and all is quiet. Consequently, the print is then heard as either a pre- or post-echo of the voice.

Print-through has been considered an inherent disadvantage of magnetic recording. However, generally the print is far enough below the sound level on the tape to cause no trouble. In fact, in nearly all home-type recorders, the noise level in the reproducing amplifier alone is far higher than the print. However, by contrast, on professional machines that employ excellent amplifiers with very little background noise, the print is sometimes audible.

One theory advanced concerning the cause of print-through is that in every magnetic material there are many "soft" particles. As we have already seen, "soft' 'magnetic materials are easily magnetized. The normal "hard" iron-oxide particles that are coated on a tape coating require a strong magnetic field before they become magnetized. The bias and record signals are many times stronger than the field that is left on the tape. Consequently, any "soft" particles in the tape are weak enough to become magnetized from the recorded signals on the adjoining tape layers.

According to one school of theorists, the weak "soft" particles line themselves up with the magnetic field on the next tape layer on the reel. It is obvious that any shaking field such as a bias or magnetic field will aid the "soft" particles in lining up. So, therefore, a stored reel of magnetic tape should be kept away from any magnetic fields. Also, a high temperature will aid "soft" magnetic particles in lining up.

It is also believed the strong bias used during the record process makes the particles susceptible to print. That is, when the tape is being recorded, the particles are sensitized and, more or less shaken up, are left in a random position out of their normal field. As a result, they are highly susceptible to magnetism from adjacent fields. Then when wound on a reel next to a turn of tape that has a highly recorded signal, the weak particles will very easily line up with the adjacent layer.

Despite lack of agreement regarding the cause of print-through, considerable information has already been accumulated. It is interesting to note that during print the magnetic particles line up very rapidly at first because of their highly sensitized position. But, as time goes on, the rate of print becomes progressively less.

The signal-to-print (SP) ratio is the ratio of the maximum recorded signal, often assumed to be that signal which produces 3% harmonic distortion, compared to the print or echo level.

If one were to make a recording, then play it back in a matter of several seconds after it was recorded, let us assume a 50-dB signal-to-print ratio would be noted. However, if five minutes were allowed to lapse before playing the tape, a 48-dB signal-to-print ratio would be present. If one were to wait an hour, the signal-to-print ratio would be 47 dB and, after 5 hours, 46 dB. Following 8 hours, very little more happens to the signal-to-print ratio. Typically, it would settle down to a value of approximately 45 dB and never become worse.

However, if the reel is played several times through a recorder, it is obvious that the print layers will not line up in precisely the same pattern. That is, with each tape winding, the area which has received the print will not line up exactly adjacent to the tape area that first made the print. Then, each time the tape is played, the same signal source will print in a slightly different area. Consequently, the print will eventually become noise. Although obvious, few people realize that if a tape with a high signal-to-print ratio is played several times, print no longer remains print but turns into noise and is heard as hiss.

New resin binders make possible the production of low-print tape with an output comparable to the best general-purpose tapes. With conventional binders low print-through can only be achieved by sacrificing output.

SPLICING PROBLEMS AND TECHNIQUES

One of the most appealing and useful characteristics of magnetic tape is the ease with which it may be cut, edited, and respliced without impairing the quality of the recorded information on the tape. The fact that tape ends can be joined so easily has led to careless techniques which have resulted, inevitably, in considerable trouble in recorder operation.

At least six months are required to properly train a good tape editor. Considerable skill is demanded in professional recording studios for which there is no substitute for experience. However, this does not mean the rank amateur cannot make acceptable splices. Provided a few simple rules and precautions are followed, the novice can make splices which will be as strong as the magnetic tape itself—splices which will wear without fraying or loosening and which can be detected only by the most sensitive computer-type recorders insofar as magnetic performance is concerned.

Most pressure-sensitive tapes now on the market have adhesives which are unsatisfactory for use in splicing magnetic tape. Splices in the wound roll of tape are subject to considerable pressure and temperature variations, and the adhesive used in ordinary pressure-sensitive tapes wil "creep" or "bleed" around the splice. This is a particularly serious condition, since not only is the strength of the splice impaired but also the adhesive invariably contaminates the magnetic side of the tape, causing adjacent layers to adhere, one to the other, resulting in the loss of the recorded information in the contaminated area due to poor head contact. The adhesive can also transfer to the heads and guides of a recorder, and thus ruin a considerable amount of tape.

To prevent trouble from the adhesive of the tape, a specially formulated splicing tape is available, guaranteeing trouble-free performance if properly used. Translucent white in color, this acetate-backed, pressure-sensitive tape for splicing is universally used with quarter-inch magnetic tape. Its 1.5-mil thickness provides thin yet strong splices. Its pressure-sensitive adhesive has these thermosetting characteristics: The bond with the magnetic tape increases under the effects of time and temperatures. Also, the adhesive will not ooze or bleed around the edges of the splice. The splicing tape eliminates any tendency to gum up recording heads or cause adjacent layers of tape to stick together on the reel. This standard splicing tape is available in standard rolls 7⁄32″, ½″, and ¾″ in width by 66′ or 150″ in length.

Apart from considerations of strength, the excellence of a splice in magnetic-tape editing can best be judged by the degree of disturbance it causes during playback. The use of the right splicing-tape material, together with proper splicing methods, can produce splices which are essentially "acoustically invisible."

In the quarter-inch magnetic-tape field, the diagonal butt splice is by now a tradition. Such splices, properly made, wear without fraying or loosening. To make a perfect splice the ends of the tape to be joined should be held in some rigid fashion to prevent displacement while the splice is being made. There are three ways of doing this:

1. The most common method, even among professional tape editors, is to hold the tape securely between fingers, cutting with a pair of scissors.
2. By laying on a flat surface (example, soft wood block or sheet of rubber) and cutting with a razor blade. This technique can be further improved by placing a straight edge along one side to position both tapes.
3. By using a commercially available splicing block. Fig. 10-4 shows some that are available for use by home hobbyists. A professional-type splicer for magnetic tape is also shown in Fig. 10-4.

The two ends of the magnetic tape to be spliced should be overlapped sufficiently to enable easy cutting and true alignment of both tape ends, as shown in A of Fig. 10-5, providing the recorded information will allow this much tape to be removed. Naturally, no cutting will be done at all if a break in the tape is being repaired.

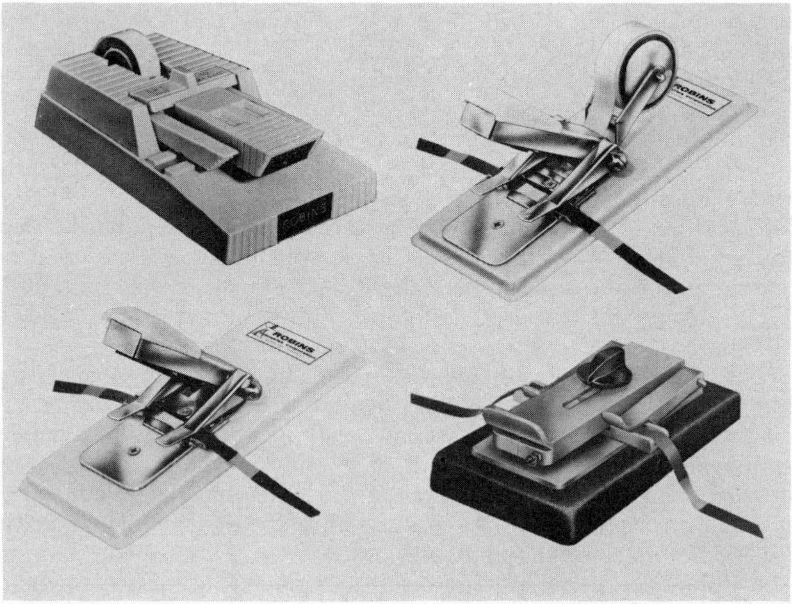

Fig. 10-4. Standard splicing blocks used by amateurs and professionals.

Line up the ends of the tape so both sections are butted and are in alignment with each other, as shown in B of Fig. 10-5. The splicing tape must always be placed on the backing (or "shiny") side of the magnetic tape.

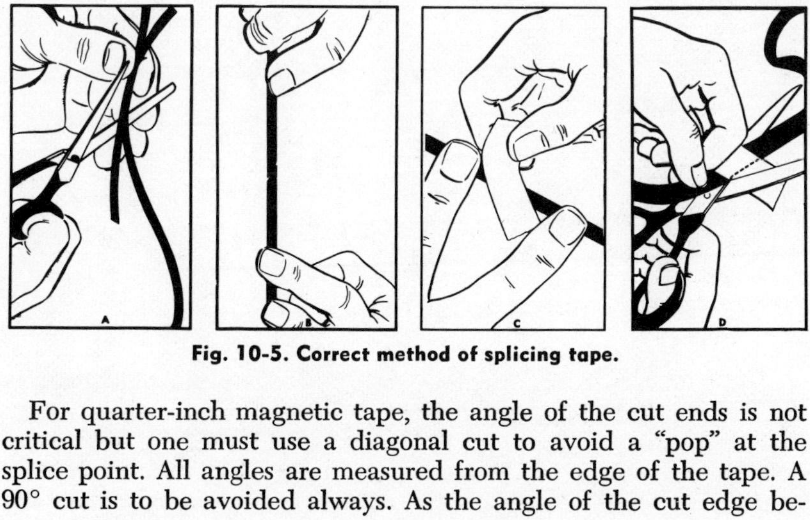

Fig. 10-5. Correct method of splicing tape.

For quarter-inch magnetic tape, the angle of the cut ends is not critical but one must use a diagonal cut to avoid a "pop" at the splice point. All angles are measured from the edge of the tape. A 90° cut is to be avoided always. As the angle of the cut edge becomes smaller, the strength, flexibility, and magnetic invisibility of the splice become greater. A 45° angle is satisfactory although a 30° splice is approximately as strong and flexible.

A piece of standard splicing tape is centered over the butted ends parallel to the splice as shown in C of Fig. 10-5. The width of the splicing tape used should never exceed ¾″ for general work, since wider tapes impair flexibility. After firmly pressing the splicing tape into position, rub firmly with the fingernail or other semihard object to press out all air bubbles. The excess splicing tape is trimmed by cutting into the magnetic tape slightly as shown in D of Fig. 10-5. Thus danger of exposed adhesive from the splicing tape is eliminated.

The techniques of splicing magnetic film in the motion-picture industry differ widely from the procedures followed in editing ¼″ magnetic tape. Because of sprocket holes and the stiffer backing employed in the manufacture of magnetic tape film, rigidity is increased. Consequently, the edges of the splice have a tendency not to conform as readily when passing through the head, tending to lift at the edges of the splice.

A different type of splicing tape is employed in editing magnetic film. While identical in adhesive characteristics to standard splicing tape, the tape is cut to 35 mm, 17.5 mm, and 16 mm widths. Also translucent white, it is available with either double or single perforations for standard sprocket holes. No thicker than regular splicing tape, it is made with a tough polyester backing.

CHAPTER **11**

Production and Test Procedures

In order to evaluate new equipment and to localize and run down bothersome recording defects, a knowledge of testing techniques is essential. To properly test magnetic tape and tape recorders, a series of test instruments is required.

The equipment need not be of expensive laboratory caliber such as that shown in Fig. 11-1. All test instruments required are relatively inexpensive and can conveniently be kept on hand in recording studios and broadcast stations. In addition, the advanced high-fidelity enthusiast would do well to consider the addition of several test instruments to his gear in order to keep his recorder in peak operating condition.

USEFUL TEST INSTRUMENTS

Among the most useful test instruments is an audio oscillator capable of reproducing the audio range of frequencies (Fig. 11-2). A good quality audio oscillator is not expensive. There are audio oscillators available in kit form for as low as $35.

An output meter such as shown in Fig. 11-3 is also on the "must" list of necessary test instruments. Most volt-ohmmeters have a dB scale which serves satisfactorily. The output meter should read essentially flat over the audio range and can be calibrated against the audio oscillator. If the output meter reads 2 dB higher at 10,000 Hz than at 1000 Hz, the 2 dB should be subtracted in frequency runs. The error in the meter must also be taken into consideration when making measurements.

In addition to an audio oscillator and an ac meter for reading output, a measuring device is needed to determine distortion. Several excellent intermodulation distortion measuring devices selling in kit form are available for slightly over $50. Other precision intermodulation measuring instruments of laboratory quality sell at upwards of $500. However, it is possible to make highly accurate harmonic-distortion measurements with a filter selling for approximately $20.

In the distortion tests to be discussed later in the chapter, it is assumed that the filter method of measuring will be used. A filter can be used in two ways. First, the filter can be used to reject the signal being recorded, thus passing only the harmonic products. Or, second, the filter can be used to pass only the desired harmonic product, rejecting all other frequencies. In other words, one may use a filter as a rejection device, rejecting the signal being recorded and passing all harmonics, thus reading total harmonic distortion. A filter may serve as a pass filter, used to pass only the harmonics being measured, enabling one to check second- and third-harmonic distortion independently.

TESTING MAGNETIC TAPE

In testing a magnetic-tape recorder, one is, in reality, testing two things: the magnetic recording tape and the recorder itself. Mag-

Fig. 11-1. Typical commercial testing installation.

netic tape is an inseparable part of the recording process. Therefore, in order to realistically test a recorder, it is essential to understand and evaluate certain important characteristics of the tape.

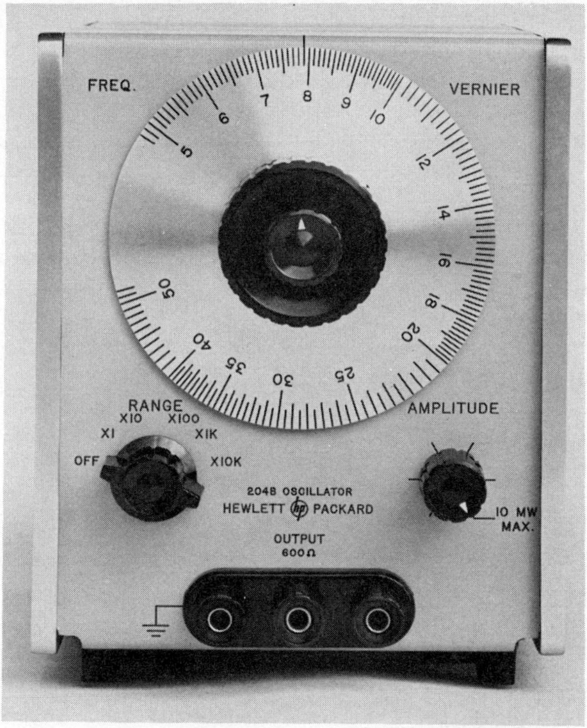

Courtesy Hewlett-Packard Co.

Fig. 11-2. High-quality audio oscillator.

When combined in finished form, the backing, the magnetic-oxide coating, the binder, the plasticizer, and the lubrication produce a magnetic tape. In testing a magnetic tape, all tape properties can be divided into two main classifications: (1) physical properties of the tape and (2) magnetic properties of the tape. In properly evaluating the performance of magnetic tape, all physical and magnetic properties must be considered since each affects performance of a tape.

PHYSICAL CHARACTERISTICS OF MAGNETIC TAPE

The physical characteristics of magnetic recording tape must be evaluated in any testing procedure. It is important to know the strength of a particular backing material. Under an applied tensile force, what is its break point? What is its yield point?

Tape should have a reasonable strength so that it will not break easily. However, practically all tape breakage is caused by nicks on the tape edges rather than the tape being pulled apart. The tensile strength of most acetate-backed tapes is approximately 5.5 pounds, while the operating tension exerted by a recorder is seldom more than a few ounces. However, since the strength of acetate is equivalent to its edge strength, any nick or tear on the tape's edge represents a potential break point. The nicks generally occur while threading the tape into the recorder. Care should be taken to place the tape properly in the recording mechanism.

Another important property of magnetic tape is that the oxide should adhere firmly to the base and not rub or flake off. Some tape constructions have been known to become brittle and flake off with the passage of time. Most present-day tapes, however, have been greatly improved in this respect.

Other factors that affect tape performance are cupping and curling. When the tape is subjected to high temperature and humidity,

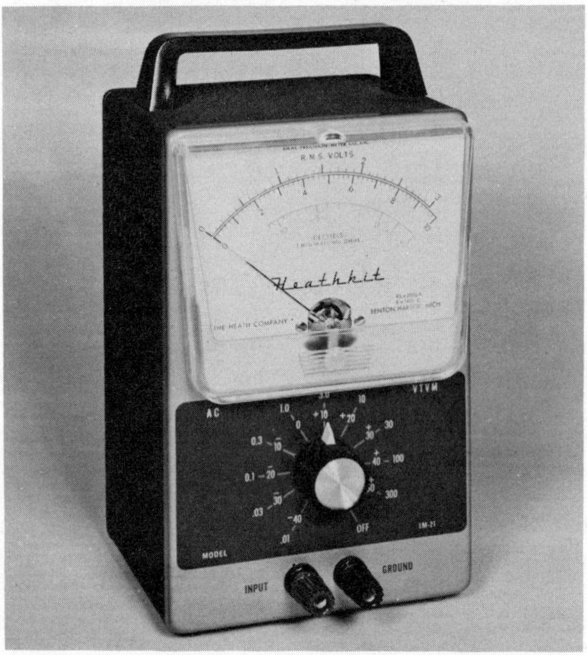

Courtesy Heathkit

Fig. 11-3. Kit-type ac vtvm.

will the tape remain flat or will it tend to curl or cup? Poor tape under high humidity conditions will tend to curl, thus making poor contact with the recording head. Poor head contact results in the loss of high frequencies and serious amplitude variations.

What is the layer-to-layer adhesion of a tape? Can one turn of the tape be freely pulled from the next or will the layers tend to stick, indicating an adhesive, sticky binder material or excessive smoothness of the surface?

A "tacky" tape will tend to move past either the tape guides or the head in a high frequency series of jerks that emit an audible squeal. While the audible squeal is not itself recorded, because it is mechanical in nature, it affects the recording, probably causing distortion in addition to an audible squeal effect.

Tests of Physical Properties

The following tests are typical of those used in testing laboratories in evaluating tape construction. Each test is but one method (although in all cases not necessarily the best way) of measuring a specific tape property.

The Backing Material—The backing material is determined by visual inspection and chemical analysis. The strength of the material, stability, and smoothness are then determined.

Tensile Strength—The tensile properties can be measured with a Scott inclined-plane tensile tester (Fig. 11-4). The result is the tensile stress in pounds required to break the tape.

Blocking—The test for layer-to-layer adhesion (blocking) consists of tightly winding several layers of tape on a stainless-steel mandrel and subjecting this sample to a 16-hour conditioning at 130° F, 90% relative humidity. The tape is then subjected to a 4-hour cycle at 130° F and 0% relative humidity and finally allowed to return to room conditions. When cooled, the tape is unwound and inspected for adhesion between layers and transfer of the oxide coating to adjacent layers. This test simulates the effect of natural aging and humidity cycling on layer-to-layer adhesion of the tape.

Curling—In the curling test, small samples of the tape are placed in each of two chambers having relative humidities of 0% and 85%. The samples are viewed "end on" and the angle of curvature is accurately measured.

Anchorage—To test for coating anchorage to the base, the oxide coating is stripped off with pressure-sensitive adhesive tape. The degree of removal indicates the firmness of the bond between the coating and the backing. Another simple test for oxide anchorage is to pinch the tape between the fingers and roll it sharply, indicating whether the oxide has good adhesion to the base.

Caliper—Tape thickness is measured on special calipering devices which measure accurately to the hundred-thousandth of an inch.

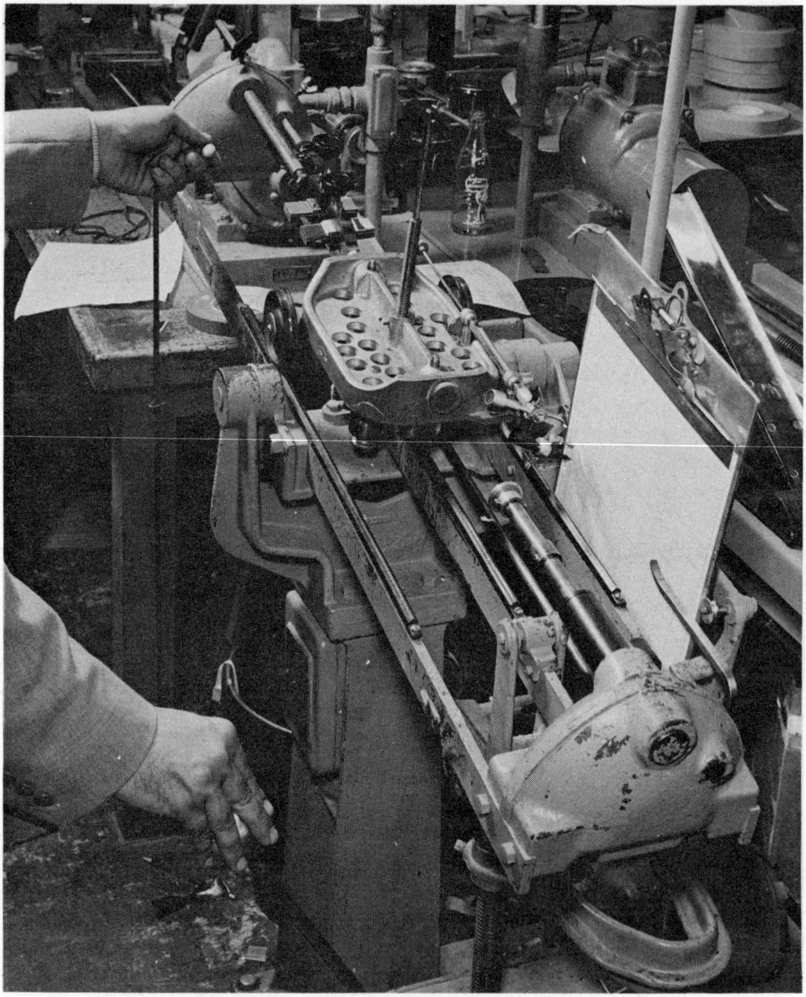

Fig. 11-4. Scott inclined-plane tester used to test tensile strength of tape.

Evaluation of Magnetic Properties

The following are the magnetic properties of a tape and how they can be measured and evaluated. Some of these properties can only be tested in a well-equipped laboratory.

Output for 1% Distortion—The output of a magnetic tape is generally determined by the coating thickness or, through newly perfected techniques in magnetic tape construction, by improved formulations which increase the oxide density without adding to tape thickness.

The output test for 1% third-harmonic distortion is made on a professional recorder adjusted at peak bias for the tape under test. A 1000-Hz signal is recorded on the tape. The input signal strength is increased until the output is 1% distorted (third-harmonic). This output is the maximum obtainable before seriously overloading the tape. A high output for a given distortion is essential in obtaining a good signal-to-noise ratio.

Sensitivity—The sensitivity of a tape is a measure of how much output or signal is obtained from the tape for a given input. This is generally measured on a comparative basis. For example, tapes of two manufacturers are spliced end to end. A given signal at low or middle frequencies is applied, well below the distortion or overload region of the tape.

After measuring output at 1% third-harmonic distortion, the 1000-Hz input signal strength is reduced to a standard value, fixed at a point well below 1% distortion for all tapes. The output of a given tape for this fixed input is, then, a measure of tape sensitivity. The output is observed for each of the tapes. If one tape has 2-dB more output than the other tape, then that tape is said to be 2-dB more sensitive.

In a sense, high sensitivity may indicate "softness" (magnetically). The high coercivity tapes, such as chromium dioxide, require a stronger signal to secure a specified level, but this guarantees permanence and freedom from print-through.

In a sense, a sensitivity is desirable since it requires a minimum of recording signal power to obtain full output. It is vital that tapes have the same sensitivity, especially when the tapes are spliced together. To cite an example of the seriousness of the problem, let us assume two tapes of varying sensitivity are spliced together. One tape may be 4-dB less sensitive than the other. When recording at full band, suddenly there will be a drop of 4 dB in the volume when the less-sensitive tape passes the magnetic record-reproduce head.

Tape Distortion—Another measure of tape performance is its output for a given distortion. As explained in the chapter on amplifiers, the tape itself introduces only third-harmonic distortion. The amplifier, however, will introduce both even- and odd-harmonic components. Generally, second and third harmonics are the strongest. However, if there are any magnetized regions on the recorder such as magnetized heads, the tape will introduce second-harmonic distortion.

Using a filter measuring device, the measure of tape output for a given distortion is determined as follows: A middle- or low-frequency signal (between 500 and 1000 Hz) is recorded on the tape. The amplitude of the signal is slowly increased until playback distortion reads a fixed value of 3%. Then the output from the tape is observed at this point.

Frequency Response—Still another test of a tape's magnetic properties is its frequency response. This is, in reality, a measurement of tape sensitivity at different frequencies. For example, on a given recording mechanism one tape reproduces the same output at a high frequency, say 10,000 Hz, as at 1000 Hz. A second tape may reproduce 2-dB less output at 10,000 Hz than at 1000 Hz. The second tape is said to have 2-dB less output at 10,000 Hz when compared to the standard reference test tape.

With the input signal level set below the 1% distortion point by the amount of pre-equalization used in the recorder (input at least 15 dB below zero level for most recorders at 15 ips), the output for a 1000- and a 15,000-Hz signal is noted. When the low-frequency output in decibels is subtracted from the high-frequency output, the result represents either the loss or gain in output of the tape at high frequencies. A good output at high frequencies is a "must" for high-fidelity recordings and for machines that are operating at slow tape speeds.

Noise Level—Generally the easiest method of reading noise is to hold a strong permanent magnet against the tape. By playing the tape at a fixed gain setting of the playback amplifier, the overall dc noise is measured. The noise figure indicates the amount of modulation noise contributed by nonuniformities on the coated surface of the magnetic tape.

Uniformity—To test for uniformity, again using a 1000-Hz input signal which is held at a constant level, the recorder output is observed on a meter. Careful watch should be kept for the maximum output deviations. On present-day recording tape, the sensitivity should not vary more than one quarter decibel within a reel and not more than one half decibel from reel to reel. A more exact measurement may be made by feeding the recorder output to a magnetic-pen recorder with good response up to 50 or 100 Hz. The uniformity is then taken as the average output variation in decibels over the length of tape wound on the reel.

Coercive Force and Remanent Flux—The magnetic tests for coercive force and remanent flux are made on a dynamic hysteresis-curve tracer, a laboratory instrument. The coercive force largely determines the signal strength required to record or erase the tape. The remanent flux indicates the approximate signal output from the tape.

TESTING A RECORDER

The most common measurement made of recorder performance is the frequency-response test. This is the most common test of a recorder, perhaps, because it is the easiest test to make and is generally indicative of the overall performance of a recorder.

A good recorder will have what is known as a reasonably flat response. That is, over a given range of frequencies, the output will vary only a small amount. There will be no sudden drops of output at certain frequencies or rises in output at other frequencies. Professional recording equipment is often flat to within one or two decibels over the entire audio range from 30 to 15,000 Hz. In contrast, some of the less expensive home recorders may vary as much as 6 to 10 dB over the range of 70 to 8000 Hz.

Frequency Response

To measure frequency response, the audio oscillator is connected to the input of the recorder as shown in Fig. 11-5. The amplitude is set so that the signal level recorded on the tape is down by at least the amount of pre-equalization used in the recorder (generally 15 dB below the normal record level).

Magnetic tape overloads at approximately the same amplitude regardless of frequency. Since most recorders often boost the high frequencies during recording, the frequency run must be made at

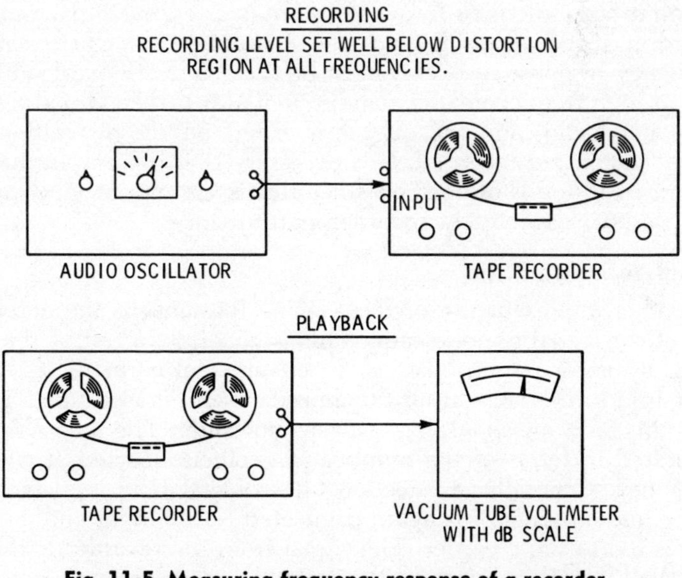

Fig. 11-5. Measuring frequency response of a recorder.

a low enough level so that the high frequencies will not be boosted to a point where they will overload the tape. If, when making a frequency-response measurement, the record level is too high, the high frequencies in the region of pre-emphasis will overload the tape and will not be reproduced properly. A tape recorder with normally excellent frequency response would then appear to have very low output at high frequencies. As shown in Fig. 11-6, the

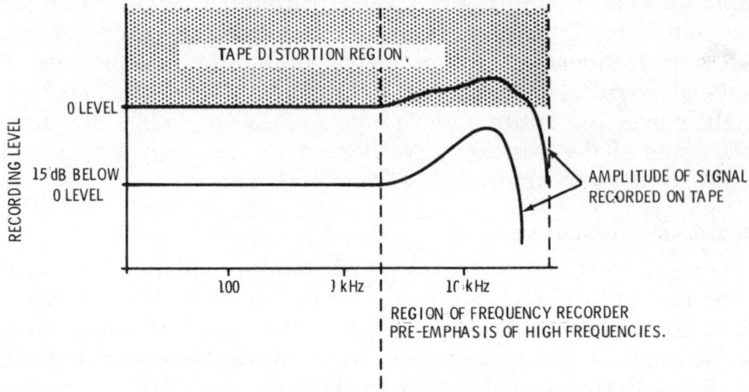

Fig. 11-6. Disadvantage of high-volume setting for frequency checks.

record level for frequency-response checks should be made at least 15 dB below "zero" level for most machines. That is, if in the equalization process the high frequencies are boosted by 15 dB, one must record at 15 dB below the normal recording level when making a frequency-response test. This must be done to avoid overloading the tape at high frequencies. The audio oscillator is then varied over the frequency band, and the output is noted on the ac voltmeter or output meter and recorded. It is necessary that the output meter be calibrated in terms of decibels since this is the unit of measure generally used in specifying recorder performance.

Signal-to-Noise Ratio

Another important test of recorder performance is the determination of the signal-to-noise ratio. This is generally a test of the maximum distorted output, taken as 3 percent total harmonic distortion, compared to the noise from the amplifier with erased tape. To perform this test, one should have a rejection filter. The filter should be calibrated in terms of the number of decibels rejected at a certain frequency. Generally, a rejection filter of 400 Hz is satisfactory.

The audio oscillator is again connected to the input and a 400-Hz tone is fed to the recorder. The signal from the recorder is then fed into the filter, thus rejecting the 400-Hz or fundamental frequency.

The output from the filter then represents the harmonic products. However, care should be taken to subtract any fundamental that is passed by the filter.

The tape output is then set so the filtered output from playback represents 3 percent harmonic distortion. The output at this point is then observed. Next, cleanly erased tape is played back on the same machine, and the noise is noted. The difference in output between the 3 percent point and the noise level is the signal-to-noise ratio of the recorder. Professional recording equipment often has a 60-dB or better signal-to-noise ratio. On home-type recorders, a figure of 40 dB may be considered good.

Flutter and Wow

Still another test of a recorder is for flutter and wow. To accurately express the amount of flutter and wow in a percentage generally requires an expensive flutter and wow meter. However, one can get along without this costly laboratory measurement device. The human ear is most sensitive to flutter and wow around three thousand Hz. Therefore, to measure flutter and wow, it is well to record a signal of approximately 3000 Hz and then to play it back, listening carefully. Any flutter and wow at 3000 Hz can be easily discerned. With a little practice, one can quickly judge what is acceptable and unacceptable. Generally, if only a slight amount of flutter and wow can be discerned, the recorder is operating properly. If the amount is noticeable or irritating, chances are that the recorder has an excessive amount of flutter and wow.

Distortion

To measure distortion in a recorder the output is passed through a rejection filter, and the distortion products are read for a given output level. The equipment required for distortion measurement is an audio oscillator with good waveshape, a vacuum-tube voltmeter capable of reading over a fairly wide range of voltage, and a bandpass filter. A block diagram of this method is shown in Fig. 11-7. Since it is customary to measure distortion at 400 Hz, a 1200-Hz bandpass filter is ideal for measuring third-harmonic dis-

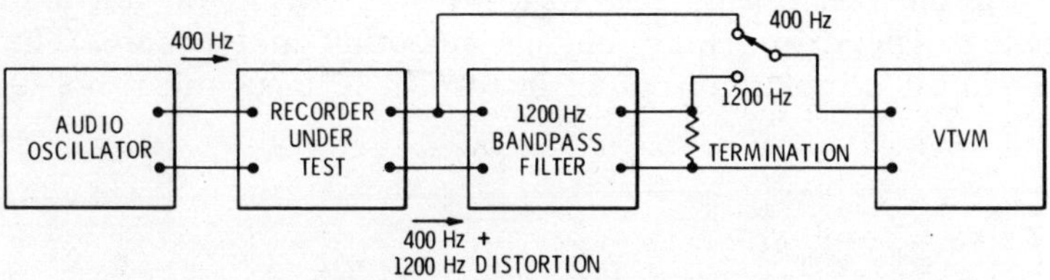

Fig. 11-7. Distortion-measuring setup.

tortion. The filter should have a rejection of at least 60 dB at the fundamental test frequency (400 Hz) if highest accuracy is to be obtained.

Before making the test it is necessary to calibrate the system to take into account the insertion loss of the filter. Since the input determination will affect this value, it is best to calibrate the filter from the actual recorder under test. To do this, the filter is disconnected and the output level of the recorder checked at 400 Hz and at 1200 Hz to determine if it is the same at these frequencies. If not, the input to the recorder must be readjusted at one of the frequencies to compensate for this discrepancy. The filter is then connected to the recorder and a level reading taken at 400 Hz at the input to the filter. The oscillator is then set at 1200 Hz and the level is read at the output of the filter. (The input level to the recorder must be readjusted if necessary as previously determined.) The difference between these readings in decibels is the insertion loss of the filter.

In making a distortion-measurement test, a 400-Hz signal is fed through the recorder and level readings taken at both the input and output of the filter. The difference between these readings in decibels minus the insertion loss of the filter is the true ratio between the signal and the third audio-harmonic component. This can be converted to percent by reference to the alignment chart given in Fig. 11-8.

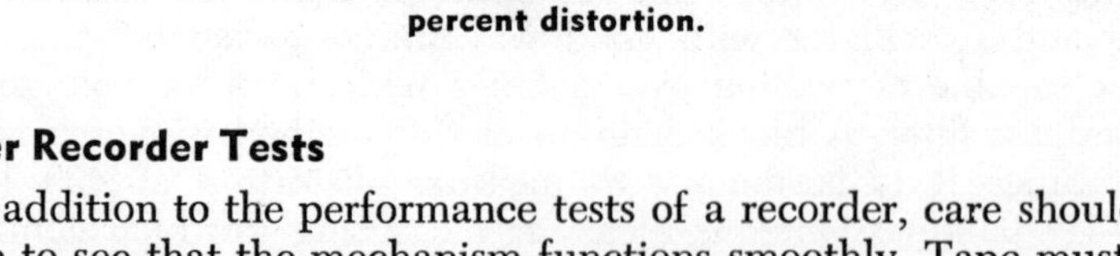

Fig. 11-8. An alignment chart for converting decibels (dB) to percent distortion.

Other Recorder Tests

In addition to the performance tests of a recorder, care should be taken to see that the mechanism functions smoothly. Tape must not spill off the reel during fast forward and rewind or starting and stopping. The drives must engage smoothly and evenly. The recorder must be relatively quiet in respect to mechanical noise.

CHAPTER 12

Specialized Applications

The tape-recorder enthusiast, like all hobbyists, is seldom satisfied until he has acquired at least some of the many accessories that are available. The tape recorder, much like the camera and the disc player in the past, has become the newest entertainment facility for the family. Actually, the tape recorder is less trouble than a camera, because magnetic tape, unlike photographic film, never has to be sent out for developing.

There are hundreds of different things that can be done with a tape recorder. The good quality tape recorder is a fine instrument capable of reproducing the full spectrum of sound. As a creative instrument, the tape recorder is an essential tool for the author, the actor, or the musician. The tape recorder serves in many ways—from language learning to private stereo-headset music listening.

STUDENT RECORDERS

Tape recorders have been designed especially for student use. The Ampex E-65, shown in Fig. 12-1, is engineered specifically to take the abuse of daily student operation and provides simplified student controls. The electronic components are designed for 50,000 hours of service—more than 5 times the life of ordinary home recorders. It is particularly suited for use in language instruction. In addition to student use, the E-65 can be used by instructors for recording master tapes, either in the classroom or at home.

An elaborate language-lab installation by Crown International is shown in Fig. 12-2. In this installation, a total of 60 tape transports are included in the rack. An installation of this type can be made with the machines installed back-to-back, 30 machines on each side, in a space only 28 inches wide and 12 feet long.

An instructor's console for use with this system is pictured in Fig. 12-3. One instructor can operate up to 30 machines at once. All the components for a complete language lab are pictured in Fig. 12-4. The six small units at the front of the desk are the student control boxes.

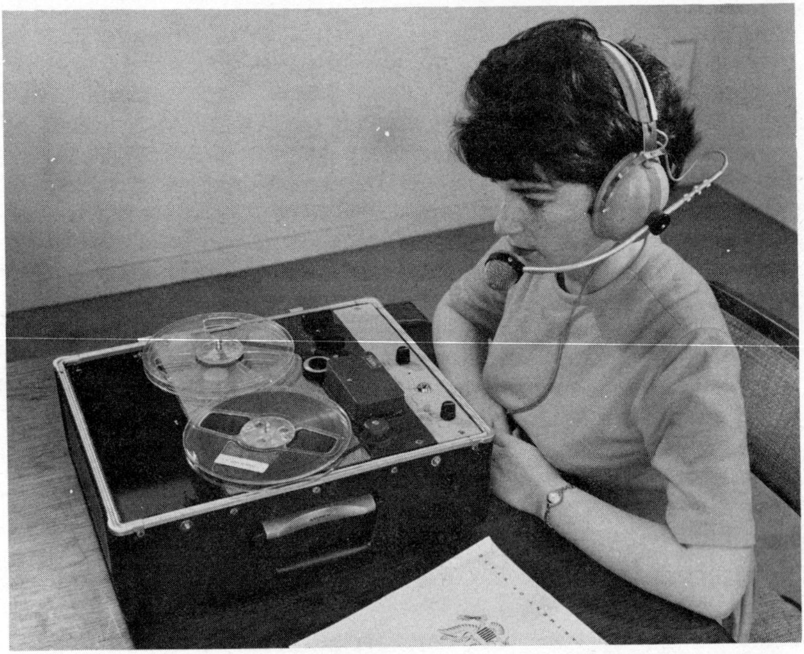

Courtesy Ampex Audio Inc.

Fig. 12-1. Recorder specifically designed for student use.

Photocell circuits are employed in the Crown International machines for a wide variety of automatic operations. In this system a photocell is illuminated by a No. 47 pilot light whenever a section of clear tape (or no tape) falls between the light and the photocell. Using this system, it is a simple procedure to "mark" a point on the tape to activate a special function. These "windows" in the tape may be used for the obvious purpose of automatically shutting off the machine in case of tape breakage or at the end of the tape, but they can also be used to operate two or three transports in sequence. Using the windows, a section of the tape can be automatically repeated or the tape can be automatically rewound and cued up so it can by played again. In another application, windows can be employed so that a specific segment of the recording can be selected by a push button, and the machine will automatically search through the tape for the proper location. In remote operation by students in language labs, the photocell circuit prevents the student from running the tape off the end of the reel, while still allowing normal control.

Courtesy International Radio & Electronics Corp.

Fig. 12-2. An extensive language-lab installation.

Other Crown International recorders provide up to 16 hours of background music and then repeat the entire program. The R-O-M accessory provides for continuous repetition of any length tape (up to several hours) and allows for a message to begin upon command.

Many other student systems have been developed and are being improved continuously. A number of manufacturers have cassette recorders that look much like the ordinary home-use model but which include control features for audio-visual use. A common arrangement is to use the second track—instead of a "second side"—to carry control pulses. A 1000-Hz pulse changes the picture by sending a pulse to a carousel slide projector or other form of visual presentation, while a 3000-Hz pulse stops the recorder for the student to do an assignment.

Instructors have found that when the recorded audio tells the student to stop the recorder and do something, often the student will not bother but will let the recorder continue playing to see what comes next. Automatically stopping the recorder encourages the student to perform the suggested action; he then restarts the recorder by pushing a restart button.

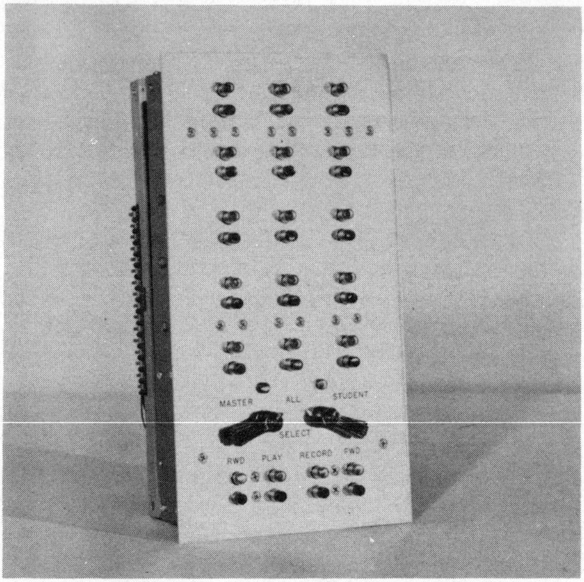

Courtesy International Radio & Electronics Corp.

Fig. 12-3. An instructor's control console.

These systems are very easy to use and an inexpensive cassette recorder can usually be used to program the tape. After the tape is programmed, knockouts prevent the student from tampering with it.

Cassette recorders are made in an almost endless variety of forms. Fig. 12-5 shows the chassis of a recorder with the cover removed that uses a precision die-cast frame for rugged use. Note that the push-button "keys" have braille marks to enable a blind person to use the machine.

Dubbing is a trade term used to describe the process of copying original tapes. It allows the recordist to make one copy or multiple copies of a tape which can then be played as originally recorded or edited to his satisfaction.

For a well-done copy, playback and recording speed must be correct and wow- and flutter-free, or the copy will not be the same as the original.

One approach is to lock the movement of both machines together by using capstans on the same shaft to drive all the tapes, both the one being copied and those being recorded (Fig. 12-6). In theory, if this is done, a little speed variation is permissible because the tape speeds will all vary together. However, designing such a machine—and using it—can be cumbersome and difficult since everything must be built around the capstan shaft. For this reason, most manufacturers design copying equipment on the principle of synchronization, operating the machines at the same speed rather than trying to lock their movements together.

Courtesy International Radio & Electronics Corp.

Fig. 12-4. The complete language lab.

The best approach to use for copying depends to a large extent on the scope of your production operation. For an occasional copy, made on a one-to-one basis, two ordinary machines coupled together, one in playback and one in record, will adequately copy a program through them at normal speed; in other words it will take an hour to copy a sixty-minute program.

For any quantity of copying, this method is impossible. One hundred separate copies, each 1 hour long, would take 100 hours to make, not counting time to load and unload between runs. One hundred copies of the same tape could also be made by connecting one machine in playback with 10 machines in record; this would accomplish the job in a little over 10 hours. Or with 100 machines available to record, the job could be done in 1 hour. Even so, the amount of time and number of machines required makes these alternatives impractical.

For short runs, a normal procedure is to use a machine that produces, say, 3 copies from a single original, running at 4 or 8 times normal playing speed. Fig. 12-7 shows such a machine for cassette duplication that will make 3 copies at one time from an original. An hour tape would take 7½ to 15 minutes to make 3 copies. Larger machines can make more copies at a time.

Courtesy Telex Communications Division

Fig. 12-5. A particularly rugged cassette machine that also includes provision for operation by a blind person.

For longer runs or for-reel-to-reel copying, the master-slave system (Fig. 12-8) is better adapted. This is obviously adapted for reel-to-reel. For producing cassette copies in quantity, the slaves are loaded with special reels of tape 0.150-inch wide, instead of the usual 0.25 inch. The same program is then recorded over and over again with a short gap between the ends of the programs. Later the tape is cut into lengths and loaded in cassettes.

A similar method can be used with cartridges, using the regular 0.25-inch-wide tape. For systems working with quantities, the speed may be 16 or 32 times normal playing speed (for example, 240 ips).

Courtesy Robins Industries Corp.

Fig. 12-6. A duplicating capstan drive that sets all cassettes, both master and copies, against the same capstan.

An hour program is thus repeated every 2 minutes, with the requisite number of tracks, by using multiple heads. With 10 slaves, such a system can turn out 300 copies an hour. That is a production rate of about 5 a minute.

To make this possible, the master is made up in the form of an endless loop, which is usually run in a loop bin (Fig. 12-9). The ends are joined together, with just enough spare to use for lead-ins on the cut-up copies. To facilitate cutting, the tape is marked with a signal that activates sensors on the cutting machine which then snips the tape at very high speed.

Notice how this works. If the copying speed is 32 times normal playing speed, the audio range during transfer is not 20 to 15,000 Hz, but 640 to 480,000 Hz. The bias frequency for the recorders, which always uses a common supply in a copying system, is also multiplied. If the master was recorded with 80-kHz, then a corresponding bias frequency for the copy machines would be 32 times this, or 2.56 megahertz.

Because the copy system uses these much higher frequencies in transfer, its frequency response must duplicate that of the audio range, only many times higher. And the requirements of the bias frequency, usually in the region of 80 kHz, must be duplicated in the region of 2.56 megahertz.

Because it is transferred at these elevated frequencies, hum is no problem. If any hum, which must be kept to a very low level in original mastering work, creeps in during transfer at 32 times the normal speed, it is the equivalent of 60 Hz divided by 32, or about 2 hertz, which is definitely subaudio and unlikely to be reproduced by any tape recorder amplifier.

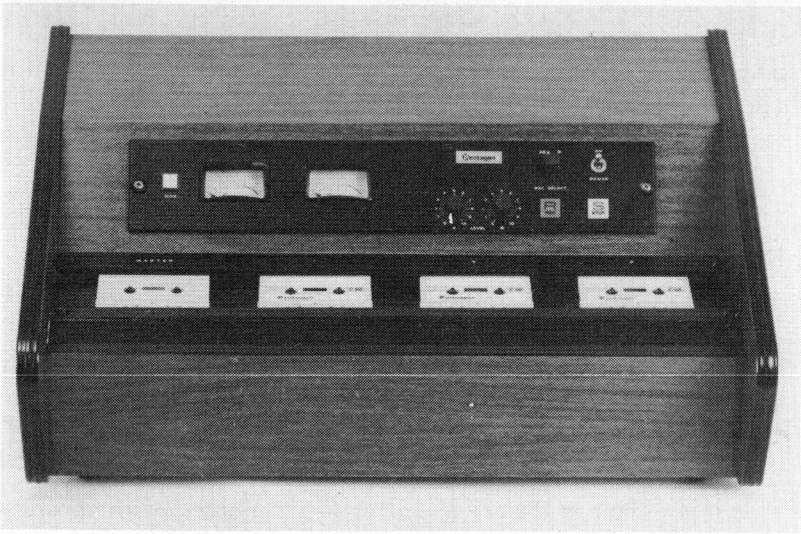

Courtesy Ampex Audio Inc.

Fig. 12-7. A small cassette master slave duplicator.

However, a 60-Hz signal can easily be used as a cutter signal. If the cutter runs the tape past a monitor head at the duplicating speed—240 ips—the lowest audio would be 640 Hz—and not often as low as that because it represents 20 Hz. Most of the audio will be very high. So a low filter can pick off a 60-Hz signal and use it to

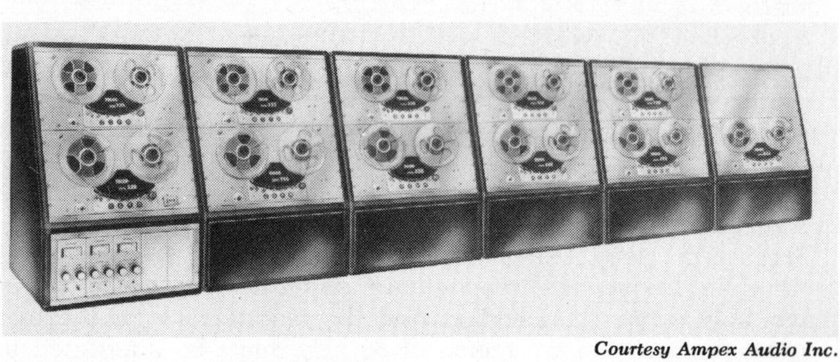

Courtesy Ampex Audio Inc.

Fig. 12-8. A reel-to-reel duplicating system used for copying reel-to-reel types.

(*A*) *Master duplicator.*

Courtesy Shure Brothers Inc.

(*B*) *Close-up of tape in bin.*

Fig. 12-9. One form of tape-loop bin.

trigger the cutter. If the cutter action is timed to snip wherever the 60 Hz occurs, it will be located in the lead-in part of the tape, as well as be at the low frequency of about 2 Hz.

Although much of this work is entirely automated in high quantity production systems, quality checks must still be performed.

The quality check system should "listen to," or at least sample, every tape produced. When many millions of listening hours are required, obviously no human listener could accomplish such a task. Electronic high-speed listeners are used to monitor certain aspects of the tapes to provide a reasonable assurance that all of the material transferred is a faithful replica and that the dubbing is functioning properly. Such a system will usually pick up any serious defect that may develop in the system and set off an alarm.

As a further safeguard, periodic spot checks are performed by human operators, mainly as a check on the automatic checkers but also to verify that minor quality defects have not crept in.

In both cartridges and cassettes, another functional check that is usually automated is one to verify that the cartridge or cassette is correctly loaded so the tape will "run" at normal operating speed. In either type, if the tape should be too tightly or too loosely spooled by the loading machines, or if a kink should develop in the

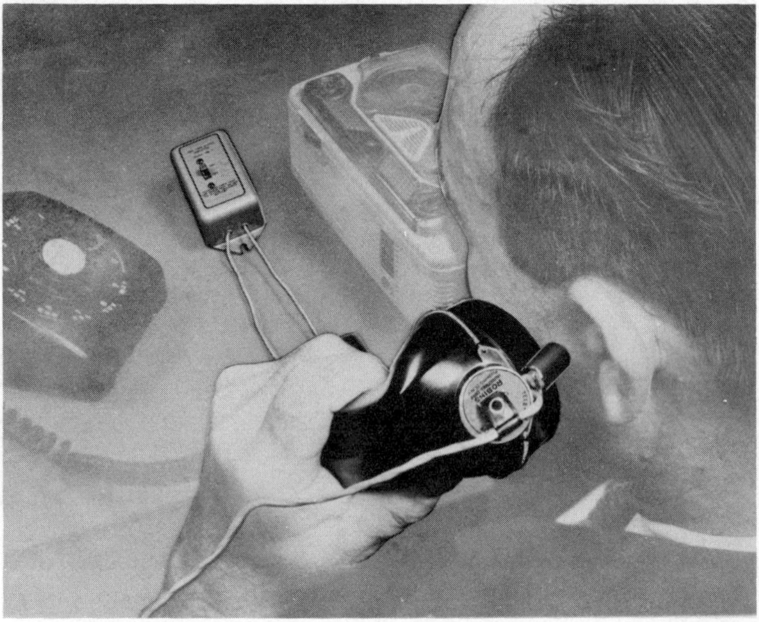

Courtesy Akai Electric

Fig. 12-10. Telephone pickup unit with beeper.

tape it could impair free running action of the package issued. This test may or may not be combined with the quality check test, depending on the way best suited to the system.

TELEPHONE RECORDERS

Recording of telephone conversations has become common practice in business and is also a convenient method of making "notes" of telephone conversations for home use. Law requires that all conversations recorded from telephone lines be identified as recorded by the application of a "beep" tone periodically during the conversation. This tone is a means of advising the person whose conversation is being recorded that it is being recorded.

Telephone pickup attachments that fasten to the telephone handset are available; they are compact, easily attached, and fulfill the legal requirements as to telephone recording. One such accessory, the Robins TRB-3 telephone pickup with a "beeper," is shown in Fig. 12-10. This accessory has a high-efficiency ferrite-core pickup coil, giving high gain and good signal-to-noise ratio. No electrical

Courtesy Akai Electric

Fig. 12-11. Communications-logging recorder.

Fig. 12-12. Cassette car recorders have many uses such as taping on-the-scene impressions of special places and events during vacation trips.

connections to the telephone lines are required; therefore, the possibility of trouble with the telephone company is eliminated.

The simple telephone pickup does a creditable job for telephone recording in a home or small office; however, a professional re-

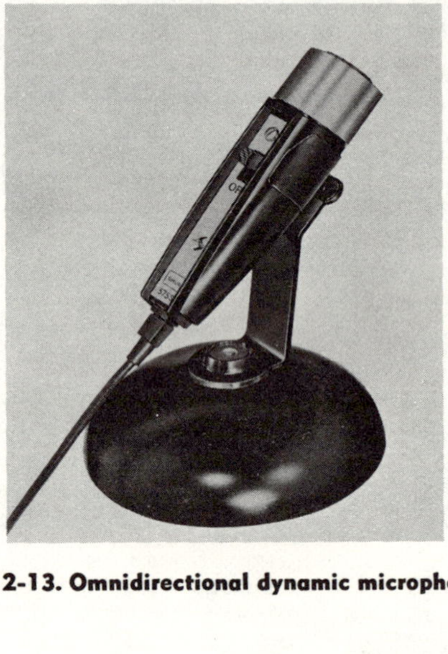

Fig. 12-13. Omnidirectional dynamic microphone.

corder is necessary for applications where traffic is heavy and results must be positive. A 2-channel slow-speed tape recorder that provides up to 8.5 hours of uninterrupted recording time for communications logging is shown in Fig. 12-11. This type of recorder keeps an accurate permanent log of voice communications. Dependable audio quality is available for police and fire departments, ambulance dispatch and report, highway authorities, military communications, airline communications, and many other functions of government and private industry.

ON-THE-GO STEREO

Installations of tape playback units and recorders in automobiles are becoming increasingly popular. These include musical and spoken recordings on either cartridges or cassettes. One such unit is shown at Fig. 12-12. These units are mounted beneath the dash and are completely transistorized. Two speakers which fill the car with stereo music are included. Deluxe home adapters are also available from the manufacturer, which allow the auto/home stereo to be utilized in the home as well as in a mobile system.

MICROPHONES

The average tape-recorder hobbyist is usually not satisfied for long with the microphone, however good it may be, that comes with his new recorder. A complete book could be filled describing the many fine microphones that are available. Ranging through a wide price range, microphones of all types suitable for tape-recorder use are offered to the amateur or professional recordist. One example of a versatile microphone that can be used either in a desk stand as a lavalier microphone or in a wall mount is shown in Fig. 12-13. The desk stand is shown in the illustration. This particular microphone is a pressure-dynamic type with a wide frequency response and omnidirectional pickup characteristics. Bidirectional and unidirectional microphones with crystal, ceramic, and ribbon-type pickups that are useful for all general and specialized recording functions are also available.

A whole variety of possibilities has opened up with the advent of 4-channel stereo, usually called quadraphonic sound. Tape is the ideal medium for this because it is easy to record 4 completely independent channels. Because there is a lot more stereo material than 4-channel material available, some modifications are necessary, so that people can play stereo on quadraphonic and vice versa.

Fig. 12-14 shows a cartridge quadraphonic recorder with a variety of capabilities.

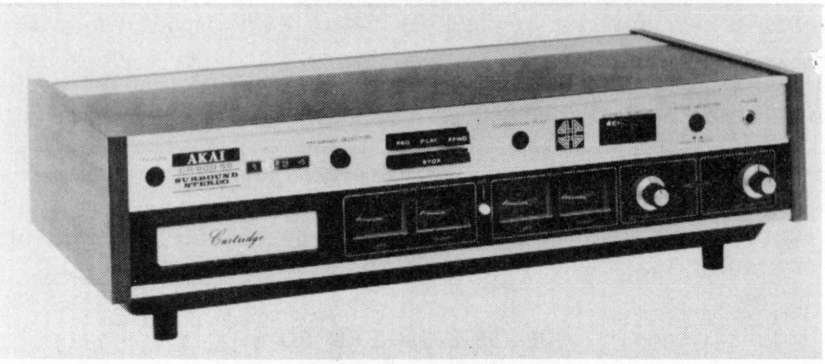

Fig. 12-14. An imported cartridge recorder with 4-channel capability and a wide range of features.

SOUND-ON-SOUND RECORDING

Sound-on-sound recording is a monophonic function which allows the performer to accompany himself over and over on a single tape to create a composite recording with almost endless variations. On recorders with this feature included, the artist can make a recording on one track and then replay the tape while recording another part of the composite recording on the second track. Theoretically, additional parts can be recorded in almost endless quantity, but practically, successive recording on the same track causes an increasing amount of deterioration in the fidelity of the recording. Extra care must be exercised in the recording operation. The use of a good microphone and special attention to the elimination of background noise will contribute greatly to the quality of the finished recording. Headphones may be used to listen to the master track while recording additional parts of the recording.

Glossary of Recording Terms

Magnetic-tape recording has developed a language of its own. Some terms have long been used in the audio field; others are common only to tape recording. Following is a listing of some of the most common terms, with definitions in nontechnical language.

Acetate base—The smooth transparent film which forms the base of most magnetic tapes.

Audible tones—Sounds which the average human ear can detect. The figure usually used is 30 to 15,000 Hz.

Automatic shutoff—Special switch which automatically stops the recorder when the tape runs out.

Bias—A high-frequency alternating current fed to the recording circuit during record to minimize distortion.

Binaural recording—Recording with two microphones on separate tape tracks, to be played back through separate amplifiers and binaural headphones.

Bulk eraser—A device for erasing an entire reel of tape in one operation.

Calendered—A trade name for tapes that have been micropolished by one process or another.

Capstan—The shaft which drives the tape and pulls it past the heads.

Cartridge—A container with a single, endless spool of ¼-inch tape, usually recorded with 4 or 8 tracks which can be monophonic, stereophonic, or quadraphonic.

Cassette—A container that houses two small spools that carry tape 0.015 inch wide and that operate at a speed of 1⅞ ips. The device was invented by the Philips Company, but it is now in wide use for an almost endless variety of purposes.

Chromium dioxide—A newer type of tape coating, capable of higher level, better quality recording and distinguished from the "regular" tapes by being black in color. Its capability may be realized only when the machine using it applies stronger bias and higher audio level in the recording process.

Counter—A device, driven by the tape-recorder mechanism, which indicates the amount of tape used.

Crossover network—Filter circuit used in multiple speaker systems which separates highs from lows and feeds them to individual speakers designed to reproduce the high or low frequencies.

Cross talk—Interference from one channel on a stereo tape with the other channel.

dB—A unit of sound measurement. Two dB is generally accepted as the smallest amount of change in loudness that can be detected by the human ear.

Deck—The tape-transporting mechanism, including the heads, drive, and tape-motion controls.

Distortion—Elements which alter the faithful reproduction of the original sound.

Dubbing—A trade term used to describe the process of copying tapes.

Dynamic microphone—Electromagnetic-type microphone which employs a moving coil in a magnetic field to produce varying voltage.

Dynamic range—The ratio between the softest and loudest sounds that a tape recorder can reproduce without distortion. Dynamic range is usually measured in dB.

Edit—Changing the sequence of a tape recording by deleting, rearranging, or adding material.

Erase head—A magnet which erases any magnetic pattern on the tape by subjecting it to a decreasing magnetic field—may be an electromagnet or permanent magnet.

Fast-forward control—The provision on a tape recorder for rapidly running tape through the machine.

Feedback—A howl or sound from a speaker caused by sound from the speaker entering a microphone connected to the same amplifier that powers the speaker.

Fidelity—A word signifying the faithfulness of sound reproduction compared to the original sound.

Flat response—The ability of a sound system to reproduce all tones in their proper proportion.

Flutter—A high-speed tone variation caused by tape moving past the head at an inconstant rate of speed.

Frequency—Number of cycles per second of a sound. Expressed in hertz (Hz).

Frequency response—The upper and lower limits of recording or reproduction of sound by a unit of audio equipment.

Gain—Amount of amplification.

Gap—The space between opposite poles of the head, usually stated in mils or microns.

Head—An electromagnetic device which consists of a ring-shaped metal core broken at the gap. Coils wound on either side of the ring energize the head during record and transform the magnetic fields to electrical currents during playback.

Head, dual-track—A head with one pole piece which covers only half the tape. The second track is recorded by flipping the tape and recording again.

Head, erase—The magnetic head used to erase the magnetic track from tape.

Head, full-track—A head with a gap covering the full width of the tape.

Head, quarter-track—A head having two pole pieces which cover the first and third quarters of the tape. The second and fourth quarters are recorded by flipping the tape and recording again.

Head alignment—Adjusting the angle of the head so that the gap is exactly perpendicular to the tape surface.

Hiss—A hissing noise originating in the amplifier or the tape.

Hum—A background noise caused by poor shielding or filtering of ac circuits.

Index counter—A counter which makes it possible to note the exact location of any particular material on the tape.

Iron oxide—The coating applied to the acetate backing in manufacturing magnetic tape.

Ips—Abbreviation for inches per second.

Jack—Receptacle for a plug connector.

Level—The volume of sound or signal.

Level indicator—A device on a tape recorder indicating the level of the signal being recorded, which serves as a warning against overrecording or underrecording.

Micropolished—A process for achieving a very fine polish on the oxide surface of the tape, enabling good contact between the oxide and the heads at all times.

Mil—One thousandth of an inch.

Monaural—A recording made on one track.

Mylar—A base material for magnetic tape. *Mylar* is stronger than acetate and is not sensitive to temperature or humidity.

Overload—More volume than the recorder can handle either on record or playback.

Patch cord—A cord used to connect the recorder to other audio equipment, such as phonographs, radios, television sets, etc.

Plug—A connector which is inserted in a jack.

Polyester backing—A plastic backing used for magnetic tape.

Power amplifier—An amplifier used to boost audio signals to sufficient strength for speaker operation.

Preamplifier—An amplifier used to raise extremely weak signals from a microphone or playback head to a usable level for the power amplifier.

Pressure pads—Small pieces of felt used to hold the tape against the heads.

Pressure roller—A rubber-tired roller which holds the tape, by spring pressure, against the capstan.

Print-through—The transfer of magnetism from one layer of tape to another layer while the tape reel is stored.

Quadraphonic—A general name for the new 4-channel sound that provides differentiation between front and back, as well as between left and right, as in stereophonic sound.

Recording noise—Noise induced by the preamplifier or other components of a tape recorder.

Reverberation—Reflection of sound in a room or cabinet, causing echo of the original sound.

Rewind control—Control for the function of rapidly rewinding tape from the takeup reel to the feed reel.

Signal-to-noise ratio—The level of the signal relative to the level of noise.

Single-track recorder—A recorder that records only one track on the tape.

Sound-on-sound—A technique of adding additional parts to a recording while the first recording is being played back. The combination of more than one part on one track.

Sound-with-sound—Adding a second track to a tape in synchronism with one track already recorded.

Spindles—The posts on which the tape reels rest on the recorder.

Splice—A joint between two pieces of tape.

Splicer—A device for making splices.

Splicing tape—An adhesive tape with a special coating which will not soften or "bleed."

Squeal—A noise made by glazed or dirty pressure pads rubbing on the tape.

Stereophonic sound—Dimensional or directional sound reproduction achieved through the use of two or more sound tracks heard simultaneously through two speakers.

Synchronous motor—A driving motor whose speed is controlled by the frequency of the current which runs it.

Takeup reel—Reel on the tape recorder which accumulates the tape as it is recorded or played.

Tape drive—The mechanism and motor which move the tape past the head.

Tape, magnetic—A high-quality plastic ribbon coated with iron oxide which retains the magnetization placed on it by the recording head.

Tape, prerecorded—Tape which has a program already recorded on it. A better term is "recorded tape."

Tape speed—The speed at which the tape moves past the head.

Track—The magnetized area laid down by the recording head on a recording tape.

Track, dual—A track that occupies slightly less than half the tape width; by flipping the tape and recording again without rewinding, a dual-track recording is made.

Track, full—A track which covers the full width of the tape.

Track, quarter—A track occupying slightly less than one quarter of the tape width; i.e., four tracks on one tape.

VU meter—A meter which indicates volume on recording or playback.

Watt—A unit of electrical power used to denote the output of speakers or the amount of current needed to operate any electrical device.

Wow—A change in sound pitch caused by variable tape speed.

Index

U

V

W